AGRICULTURE ISSUES AND POLICIES SERIES

NO-TILL FARMING: EFFECTS ON SOIL, PROS AND CONS AND POTENTIAL

AGRICULTURE ISSUES AND POLICIES SERIES

Agriculture Issues & Policies, Volume I
Alexander Berk (Editor
2001. ISBN 1-56072-947-3

Agricultural Conservation
Anthony G. Hargis (Editor)
2009. ISBN 978-1-60692-273-6

Agricultural Conservation
Anthony G. Hargis (Editor)
2009. ISBN 978-1-60876-813-4
(Online book)

Environmental Services and Agriculture
Karl T. Poston (Editor)
2009. ISBN: 978-1-60741-053-9

Weeds: Management, Economic Impacts and Biology
Rudolph V. Kingely (Editor)
2009. ISBN 978-1-60741-010-2

Soybean and Wheat Crops: Growth, Fertilization and Yield
Samuel Davies and George Evans
2009. ISBN: 978-1-60741-173-4

Effects of Liberalizing World Agricultural Trade
Henrik J. Ehrstrom (Editor)
2009. ISBN: 978-1-60741-198-7

Effects of Liberalizing World Agricultural Trade
Henrik J. Ehrstrom (Editor)
2009. ISBN: 978-1-60876-601-7
(Online book)

Hired Farmworkers: Profile and Labor Issues
Rea S. Berube (Editor)
2009. ISBN 978-1-60741-232-8

Agricultural Wastes
Geoffrey S. Ashworth and Pablo Azevedo (Editors)
2009. ISBN: 978-1-60741-305-9

Sugar Beet Crops: Growth, Fertilization & Yield
Claus T. Hertsburg
2009. ISBN: 978-1-60741-491-9

Economic Impacts of Foreign-Source Animal Disease
Jace R. Corder (Editor)
2009. ISBN: 978-1-60741-601-2

Economic Impacts of Foreign-Source Animal Disease
Jace R. Corder (Editor)
2009. ISBN: 978-1-60876-602-4
(Online book)

Essential Oils: Art, Agriculture, Science, Industry and Entrepreneurship (A Focus on the Asia-Pacific region)
Murray Hunter
2009. ISBN: 978-1-60741-865-8

Corn Crop Production: Growth, Fertilization and Yield
Arn T. Danforth
2009. ISBN: 978-1-60741-955-6

Corn Crop Production: Growth, Fertilization and Yield
Arn T. Danforth
2009. ISBN: 978-1-60876-860-8
(Online book)

Phosphate Solubilizing Microbes for Crop Improvement
Mohammad Saghir Khan and Almas Zaidi (Editors)
2009. ISBN: 978-1-60876-112-8

Organic Food - Economics and Issues
Earl D. Straub (Editor)
2009. ISBN: 978-1-60741-130-7

Global Beef Trade
Alessandro Ferrara (Editor)
2009. ISBN: 978-1-60741-121-5

Ecophysiology of Tropical Tree Crops
Fabio DeMatta (Editor)
2010. ISBN 978-1-60876-392-4

Assessing Disease Potential in U.S. Aquaculture Industry
Emilija Kalnins (Editor)
2010. ISBN: 978-1-60741-543-5

Inspection and Protection of U.S. Meat and Poultry
Finn J. Amundson (Editor)
2010. ISBN: 978-1-60741-120-8

Drivers and Restraints for Economically Efficient Farm Production
Helena Hansson Karin Larsén Bo Öhlmér
2010. ISBN: 978-1-60876-171-5

Soil Phenols
A. Muscolo and M. Sidari (Editors)
2010. ISBN: 978-1-60876-264-4

No-Till Farming: Effects on Soil, Pros and Cons and Potential
Earl T. Nardali (Editor)
2010. ISBN: 978-1-60741-402-5

AGRICULTURE ISSUES AND POLICIES SERIES

NO-TILL FARMING: EFFECTS ON SOIL, PROS AND CONS AND POTENTIAL

EARL T. NARDALI
EDITOR

Nova Science Publishers, Inc.
New York

LIBRARY OF CONGRESS CATALOGING-IN-PUBLICATION DATA

No-till farming : effects on soil, pros and cons and potential / editor, Earl T. Nardali.

p. cm.
Includes index.
ISBN 978-1-60741-402-5 (hardcover)
1. No-tillage. I. Nardali, Earl T.
S604.N58 2009
631.5'814--dc22

2009032439

Published by Nova Science Publishers, Inc. ✢ New York

CONTENTS

PREFACE

No-till farming (sometimes called zero tillage) is a way of growing crops from year to year without disturbing the soil through tillage. No-till increases the amount of water in the soil, decreases erosion, increases the amount and variety of life in and on the soil and it increases herbicide usage. There is evidence that repeated tillage destroys the soil resource base and causes adverse environmental impacts. Tillage degrades the fertility of soils, causes air and water pollution, intensifies drought stress, consumes fuel, and contributes to global warming. Today, farmers are expected to produce food in ever greater quantities. This is becoming more difficult to do in view of declining soil quality, which can be caused by soil tillage. It is becoming well known that no-till is an effective technique to reduce the degradation of soil. With this way of farming, crop residues or other organic amenities are retained on the soil surface and sowing/fertilizing is done with minimal soil disturbance. This new and important book gathers the latest research from around the globe in this field of study.

Chapter 1 - Erosion and soil compaction are two important degradation processes of modern agricultural production in Western Europe. One of the principal causes of soil degradation are deep and intensive tillage operations, especially under moist soil conditions. To minimize erosion and soil compaction, the adoption of conservation tillage or reduced tillage (RT) agriculture is widely recognized. However, crop rotations in Western Europe often include root crops that are generally assumed to be less suitable under RT agriculture because they result in a high disturbance of the soil at the formation of the ridges and at harvest. A detailed understanding of the influence of the different types of tillage on the soil properties aggregate stability, penetration resistance (PR) and field-saturated hydraulic conductivity (K_{fs}) is important for management strategies. The objective of this chapter is comparing the physical properties of silt loam soils under different tillage systems in Belgian fields.

The aggregate stability of the 0-10 cm layer was measured in the laboratory. The stability index measured by the 'dry and wet sieving' method of De Leenheer and De Boodt (1959) was 40% higher under RT than conventional tillage (CT) agriculture. The mean weight diameter (MWD) measured with the three methods of Le Bissonnais (1996) was significantly higher even after short-term RT compared to CT agriculture. The MWD after a heavy shower, a slow wetting and stirring the soil after prewetting was 19%, 38% and 34% higher for RT than CT fields, respectively. The PR was studied in the field with a hand penetrologger. The PR of fields under RT was higher in the first layers due to less (RT with cultivator or soil loosener [=RT_C]) or no disturbance (RT by direct drilling [=RT_{DD}]) of the soil structure. The crop growth of RTC fields seemed not to be affected by this superficial compaction. However, the crop growth of RTDD fields was lower than of CT fields. The PR measurements indicated a plough pan at depth in some field under CT. The measurement of the infiltration rate with a Guelph pressure infiltrometer in the field showed that Kfs tended to be lower under CT than the other types of tillage.

It can be concluded that RT_C is a form of agriculture that is more sustainable than CT agriculture in these silt loam soils. This research indicated that short term RT_{DD} agriculture under crop rotations with root crops often result in a lower crop yield, which is probably correlated with the higher PR in the 10-30 cm depth layer. Therefore, the potential of RT_{DD} agriculture in Western Europe is probably limited.

Chapter 2 - A long-term tillage experiment was designed to assess the effect of tillage frequency and intensity on rainfed grain production and quality in the semi-arid subtropical environment of central Queensland, Australia. There were four tillage treatments: traditional tillage (TT), stubble mulch tillage (SM), reduced tillage (RT) and no till (NT), each with and without applied fertiliser (N+Zn). On completion, after 20 years of treatment application, all treatments were managed using no till and appropriate fertiliser (N+Zn) application for a further 5 years.

During the 20 years when tillage treatments were being applied, the reduced tillage treatments (NT, RT and SM) outyielded TT in fewer than half of the 22 crops grown. Mean yields without fertiliser were 2.0 t/ha (TT) and 2.6 t/ha (NT) while mean yields with fertiliser were 1.9 t/ha (TT) and 2.9 t/ha (NT). During the subsequent 5 years of no till, the former reduced tillage treatments outyielded the former TT in each of the 5 crops grown. For example, the long-term NT gave an average yield of 3.3 t/ha while the short-term NT, which was formerly TT, produced only 2.1 t/ha - a 57% yield increase. This increase was due to both increased soil water storage and higher water use efficiency (WUE). Both were attributed to the development of improved soil structure, higher population densities of soil macrofauna and slightly higher soil organic carbon content. High WUE in NT was also attributed to a beneficial effect resulting from slow early growth under no till. Results indicate it takes at least 20 years to attain the full soil benefits (physical, chemical and biological) of a no-till system. The large yield responses from the three reduced tillage treatments, during and after treatment application, were realised because cropping frequency was higher than would have been appropriate for traditional tillage. Increased cropping frequency also results in higher levels of groundcover, which reduces soil erosion and creates a more sustainable farming system. A high-yielding, viable cropping system can also contribute towards environmental sustainability by reducing the need for further land clearing.

Chapter 3 - This study was conducted to identify indigenous soil erosion control measures in the Usambara Mountains, Lushoto, Tanzania. The aim was to assess the effectiveness of the indigenous soil erosion control measures in the area and their corresponding problems according to farmers' perception. The study was expected to generate new knowledge that will contribute to the efforts endeavours to link the indigenous knowledge with the recently introduced soil erosion control measures for effective soil conservation practices that are largely acceptable to the farming community. Data collection was carried out through interviews with key informants and elders in the study area, focused group discussions with community representatives, in-depth household interviews using a structured questionnaire and participant's observation during transect walks. Descriptive statistical analysis was employed in exploring the data for distribution of responses, central tendencies and dispersion. Cross tabulation and multiple response analyses was also performed to ascertain responses and percentages.The study shows that 37% of the farmer's response identified *Miraba* (Rectangular Bound Grass Strips) that do not follow contours as a unique indigenous soil erosion control practice in the Usambara Mountains. Other indigenous technologies identified include use of *Tughutu* (*Veronia subligera*) in the border of field partitions or fields (14%), and *mboji* (band of green manure) (13%). In the study area, the majority of the respondents mentioned *Miraba* (31%) as the most effective type of erosion control practice followed by contour grass strip (20%) and terraces (2%). It was observed that when Soil and Water Conservation agencies introduced contours with grass strip as a soil erosion control measure in the studied villages, they could not make distinction between *Miraba* (an indigenous technology (IT)) practiced by the farmers on one hand and grass strip on contours on the other hand as there was no critical effort made to include farmers IT in the technology development. According to elder farmers, *Miraba* were initially established to reduce the speed of the river floods runoff in the valley bottoms. However, through evolution the technology was transferred and applied on the sloping landscape where interrill, rill and gully erosion is dominant. During transect walks, severe soil erosion and gullies were observed. The interviewed key informants mentioned these erosion processes as one of the problems associated with improper terrace construction, while accelerated runoff was mentioned as the problem associated with poor implementation of the *Miraba* technique on the slopping lands. It is therefore concluded that there is a need to differentiate farmers' perception and objectives on indigenous knowledge on one hand and on the introduced technologies on the other hand in order to develop a holistic approach that include farmers IT in the overall technological development. The study underscores the importance of further research to evaluate the effectiveness of the *Miraba* technique and to improve on its performance through linkage with the recently introduced soil erosion control practices. Training on the use of introduced technologies to improve the effectiveness of indigenous technologies is strongly recommended.

Chapter 4 - Extensive cultivation of corn (*Zea mays* L.) for food, feed and biofuels may lead to a potential risk of soil degradation. Adoption of minimum tillage or no-till (NT) conservation systems help reduce this risk with benefits such as the increased residue cover, reduction of soil erosion and fuel consumption, and the improved efficiency of crop water use. All of these are important factors for sustainable crop production, especially in arid and semiarid regions. Although the production of corn under the NT system began approximately fifty years ago in the United States, and has become a more and more popular system worldwide since the 1990s, many aspects concerning the productivity improvement of conservation tillage have not been fully documented. One of the major concerns is the corn yield drag under cooler and wetter soil conditions in temperate humid regions due to the residue cover in no-till system, particularly in fine-textured clay loam or silt clay soils. It was often observed that reduction in early season nutrient uptake and early growth of the crop was associated with NT, resulting in the reduction of plant population density and grain yield. Use of starter fertilization and adoption of residue management in NT systems have been directed to alleviate the negative effects. Soil conditions of high moisture and increased rates of fertilization practices under NT are related to high N_2O emissions because they promote microbial denitrification; however it has been stated that agricultural N_2O emissions can be most effectively reduced with recommended practices to synchronize N application with crop demand. The purpose of this review is to summarize the results obtained in numerous studies carried out in different regions and to elucidate the main points of future research that allows a new approach to evaluate the impact of the fertilization management practices and the residue management measures on grain yield improvement and environment neutrality in a NT corn production system.

Chapter 5 - The investment risk of sequestering carbon as soil organic carbon (SOC) is low when compared to the susceptibility of forestry sequestration projects to disease, bushfires and droughts over a 100-year investment timeframe. As it can require decades to create a substantial carbon market asset with the slow natural accrual rates of carbon in soils, farmers will require short-term benefits to their conventional productivity from SOC additions. This chapter explores the additional effects of improving farm SOC levels on conventional farm productivity, soil nutrition, crop yields, soil water use efficiency, crop stability, disease resistance, soil biodiversity and tillage. This work also collates a broad range of management options available to farmers for utilising SOC to assist current productivity, while clarifying the risks and uncertainties of soil carbon stock fluxes, carbon markets and carbon emission policy developments in Australia.

Chapter 6 - The soil contains a significant part of the global carbon (C) stock: 3.5 % compared with to 1.7 % in the atmosphere, 8.9 % in fossil fuels, 1.0 % in biota and 84.9 % in oceans. With increasing emission of greenhouse gases (GHG) there is a growing interest in assessing the soil role as a C sink in different management conditions. The replacement of natural to agricultural ecosystems, with annual crops and based on plow tillage has resulted in decline in soil organic carbon (SOC) stocks. In tropical regions, where high temperatures and rainfall can increase the soil degradation, the use of no tillage (NT) can reverse this process, favoring the recovery of SOC, decreasing CO_2 emissions, and hence reducing the greenhouse effect. The widespread use of the NT in Brazil is characterized by sowing on to the crop residues from the previous cultivation, without any soil preparation. The dead biomass formed by the crop residues accumulated on the topsoil can help reduce erosion, temperature and moisture fluctuations at the surface, and the crops exposure to drought. These characteristics result in chemical, physical and biological alterations in soil properties affecting nutrient uptake by crops. In Brazil, the soils are dominated by variable-charge clay minerals such as kaolinite and, especially, Fe and Al oxides, which have a high flocculation capacity due to electrostatic interactions through their positive charges. The strong adsorption of SOC to active Al and Fe oxides is the cause for SOC stability in tropical soils; therefore, the interaction of organic compounds with clay in highly weathered oxidic soils can prevent SOC losses. In this context, soil C sequestration is a significant option to mitigate GHG emission in Brazilian tropical conditions

Chapter 7 - A survey was carried out to determine the effects of channel irrigation on land degradation in the vegetable growing area on the northern slopes of the Uluguru Mountains, Tanzania. Sixteen irrigation-induced landslides were characterised in terms of their dimensions, associated geomorphic characteristics, landslide typology and morphological and physical characteristics of the soils. Qualitative assessment and descriptive statistics were widely employed in the exploratory analysis of the collected data and related factors. This included the estimation of means and standard deviations of some critical variables of the study. Where applicable, Least Squares Means (LSMean) was used to compare the treatment means at 5 percent level of probability. From the results, it was observed that most of the landslides occurred on debris slopes on the linear/concave slope complex as earth flows and or earth slides with few occurring on the amphitheatres. The dominant slide scar angles ranged from 22-35°. Channel irrigation in this area provoked relatively small sized landslides with maximum width and depth ranging from 7 to 19 m and 0.3 to 0.9 m respectively while their length ranged from 6 to 70 m. The mean damaged cropland area ranged from 211 to 1327m^2 while volume of soil loss ranged from 36 to 72m^3. Both the damaged cropland area by landslides and volume of soil loss due to the latter are the major indicators of land degradation due to channel irrigation and these parameters are proposed to be one of the yardsticks in determining the sustainability of production of vegetables in the study area. Damaged cropland area was more severe on flat cultivation system compared to ridge and contour cultivation systems ($P \leq 0.05$). This phenomenon is explained in terms of loss of crop yield and reduced farmers income. The volume of soil loss followed the trend flat cultivation > contour cultivation > ridge cultivation though the values were not statistically significantly different.

Chapter 8 - The changes of the cultivation methods and primarily the domination of no-till farming in many crops seem to dramatically influence, among others, the weed populations, the relative balance and frequency of the several weed species, and consequently the weed management practices which should be adopted in order to handle each new problem. This is the case in many Mediterranean climate regions, where no-tillage agriculture progressively becomes more and more popular and assigns the future evolutions. The example of the several species of *Conyza* (*C. albida*, *C. canadensis* and *C. bonariensis*), which have already taken advantages of the no-till farming practices and consecutively cover larger areas in Greece, is typical of this developing situation. Focusing on such a case study and particularly on the several aspects of the causes, consequences, distribution and management strategies for the above-mentioned and many other weed species, could definitely offer an opportunity for understanding the overall situation and seem crucial in order to think about and invoke several constituents of integrated weed management programs on a local and global basis. The subject of this study is to review the literature on some outstanding cases of weed species and problems that have arisen due to the no-till agriculture and to concisely present the example of the *Conyza* species.

Chapter 9 - A study was conducted in the northern slopes of the Uluguru Mountains, Tanzania to investigate the effects of slope gradient, tillage depth, soil surface conditions and soil moisture on the rate of soil flux due to manual tillage. Soil flux was measured by means of an on-farm experiment in the farmers' fields. Twenty four plots (1.2 m wide and 8 m long) at different slopes: 31, 47, 51, 54, 58, 61, 65 and 67 percent were demarcated prior seasonal land preparation. Two sets of treatment at each slope site were done: deep tillage and shallow tillage under conditions of bare soil surface and soil surface with residues to both treatments. Effect of soil moisture condition was studied in areas under intensive agriculture at slopes 54 and 67 percent. Soil flux rates were monitored by Gerlarch trough method. Descriptive statistics (means, standard deviations, minimum and maximum) were widely employed in the exploratory analysis of soil flux rates and related factors. The effect of slope gradient and treatments were statistically analysed according to General Linear Model (GLM) of the Statistical Analysis System (SAS, 1999). Least Squares Means (LSMean) was used to compare the treatment means at 5 percent level of probability. The results showed that rates of soil flux due to tillage erosion generally increased with increasing slope gradient. Mean soil flux increased significantly (P<0.05) from 16 kg/m/tillage pass at slope gradient 31% to more than 60 kg/m/tillage pass at slope gradient >60%. The rates of soil flux from bare soil surfaces for both shallow and deep tillage were significantly higher (P<0.05) than soil flux from soil surface with residues. Bare soil surface promotes tillage erosion than soil surface with residues. Further analysis showed that a combination of shallow tillage and presence of residues on the soil surface can cut down the rates of soil flux by 60% compared to deep tillage with bare soil surfaces. Therefore, farmers should be discouraged to clear their fields by removing all the residues prior to cultivation. The study also revealed that soil flux increased with moisture content, a condition attributed to greater tillage depth.

Chapter 10 - Research has shown that the less mobile elements in the soil, such as P, K and Zn, become stratified in the surface 50 mm of soil after several years of continuous no-till cropping. This causes nutritional constraints to productivity when the surface soil becomes dry unless the rest of the soil profile is highly fertile or receives appropriate levels of fertiliser.

No-till crops grown in the semi-arid, subtropical environment of central Queensland, Australia, are particularly prone to nutritional disorders as a result of surface stratification. The surface soil can remain dry for long periods during crop growth while the roots obtain water stored in the heavy clay soils during the preceding fallow. Consequently nutrient deficiency symptoms have appeared despite apparently satisfactory levels of those same nutrient elements in the 0-100 mm layer, which is the standard sampling depth used to assess soil fertility.

The original aim of the 13-year experiment reported here was to assess the effect of tillage frequency and intensity, and stubble retention and removal, on soil water storage, soil nutrient status, and the growth and yield of rainfed grain sorghum in central Queensland. During the first 7 years of research, the mean yields of no till and traditional tillage (disc plough and scarifier) without fertiliser application were not significantly different. As the experiment continued, it became apparent that supplementary nutrition was required. The placement of appropriate levels of fertiliser elements (N, P, K, S and Zn) 100 mm deep and 50 mm to the side of each row resulted in large yield responses to no till. The yield of no till in comparison with traditional tillage (both with stubble retained and appropriate fertilisation) was 147%, 129% and 174% in the final three crops. These responses also reflect the outstanding potential of longer-term no till that has been demonstrated in other research work in central Queensland. Since mixing of the surface soil (by tillage) gave similar responses, it was concluded that some crop nutrients had become concentrated in the dry surface soil and were therefore inaccessible to plant roots during dry periods.

Chapter 11 - Remote sensing of canopy reflectance offers a new way to monitor deficiencies of nitrogen (N) and estimate fertilizer needs during the growth of corn (*Zea mays* L.), but traditional evaluations of this method have been primarily limited to small-plot trials. Our studies were aimed to exploring the potentials of remote sensing and yield monitoring in estimation of N fertilizer needs in production agriculture. Much of our attention was focused on the effect of application rates of N fertilizer crossed different soils in each of 12 fields. Reflectance of light from corn canopy was measured by digital analysis of aerial photographs taken in July and August. Grain yields were measured at 1-second intervals by using combines equipped with a yield monitor and GPS receiver. Responses in both canopy reflectance and grain yields to the application of N fertilizer were obtained from paired grids in adjacent strips within each field, and cumulative frequency distribution curves for the responses were generated accordingly. The observed non-linearity between relative reflectance in July and final yield responses, by using median values for individual fields, suggests that symptoms of mild N deficiencies appear much later than severe N deficiencies, hence limiting the ability of remote sensing for an early diagnosis of minor N deficiencies. The practical value of monitoring N deficiency symptoms seems to be restricted to detecting severe N deficiencies early enough to apply corrective treatments and gathering information that can be used to fertilize subsequent crops. However, the unique ability of this approach to make many observations for a small cost helps reduce errors of measurement and make estimates of fertilizer needs that can be easily evaluated and improved.

In: No-Till Farming
Editor: Earl T. Nardali

ISBN: 978-1-60741-402-5

Chapter 1

INFLUENCE OF TILLAGE SYSTEMS ON SOIL STRUCTURAL PROPERTIES

Maria Casamitjana Causa**, *Karoline D'Haene*[1], *Juan Carlos Loaiza Usuga*[2], *Donald Gabriels*[3] and *Stefaan De Neve*[3]

Albera Natural Parc, Natural Parcs Division, Catalonia Enviroment Departement, Carrer Amadeu Sudrià 3, E-17753, Espolla, Spain

[1]ILVO Social Sciences, Burg. Van Gansberghelaan 115 box 2, B-9820 Merelbeke, Belgium

[2]Laboratory of Hydrology and Soil Conservation, Forestry and Technology Centre of Catalonia, Pujada del Seminari s/n, E-25280 Solsona, Spain

[3]Department of Soil Management, Faculty of Bioscience Engineering, Ghent University, Coupure 653, B-9000 Ghent, Belgium

ABSTRACT

Erosion and soil compaction are two important degradation processes of modern agricultural production in Western Europe. One of the principal causes of soil degradation are deep and intensive tillage operations, especially under moist soil conditions. To minimize erosion and soil compaction, the adoption of conservation tillage or reduced tillage (RT) agriculture is widely recognized. However, crop rotations in Western Europe often include root crops that are generally assumed to be less suitable under RT agriculture because they result in a high disturbance of the soil at the formation of the ridges and at harvest. A detailed understanding of the influence of the different types of tillage on the soil properties aggregate stability, penetration resistance (PR) and field-saturated hydraulic conductivity (K_{fs}) is important for management strategies. The objective of this chapter is comparing the physical properties of silt loam soils under different tillage systems in Belgian fields.

* Corresponding author: Maria Casamitjana i Causa, Puig Salner 14, Castello D'Empuries, Provincia de Girona, E-17486, Spain (email: casmi242@hotmail.com, Tel. +34-972250142, Fax +34-972514339)

The aggregate stability of the 0-10 cm layer was measured in the laboratory. The stability index measured by the 'dry and wet sieving' method of De Leenheer and De Boodt (1959) was 40% higher under RT than conventional tillage (CT) agriculture. The mean weight diameter (MWD) measured with the three methods of Le Bissonnais (1996) was significantly higher even after short-term RT compared to CT agriculture. The MWD after a heavy shower, a slow wetting and stirring the soil after prewetting was 19%, 38% and 34% higher for RT than CT fields, respectively. The PR was studied in the field with a hand penetrologger. The PR of fields under RT was higher in the first layers due to less (RT with cultivator or soil loosener [= RT_C]) or no disturbance (RT by direct drilling [= RT_{DD}]) of the soil structure. The crop growth of RT_C fields seemed not to be affected by this superficial compaction. However, the crop growth of RT_{DD} fields was lower than of CT fields. The PR measurements indicated a plough pan at depth in some field under CT. The measurement of the infiltration rate with a Guelph pressure infiltrometer in the field showed that Kfs tended to be lower under CT than the other types of tillage.

We can conclude that RT_C is a form of agriculture that is more sustainable than CT agriculture in these silt loam soils. This research indicated that short term RT_{DD} agriculture under crop rotations with root crops often result in a lower crop yield, which is probably correlated with the higher PR in the 10-30 cm depth layer. Therefore, the potential of RT_{DD} agriculture in Western Europe is probably limited.

Keywords: Tillage Systems, Soil Structure, Soil Compaction, Penetration resistance, Infiltration rate

1. INTRODUCTION

The explosive population growth in the 20th century resulted in a high food demand in industrialised societies. Therefore the arable as well as animal production needed to increase through an intensification of agriculture. Until recently, modern agriculture was based on mechanization, intensive use of agrochemicals and organic manure and was focused on maximum food production without considering the long term impact on soil fertility or environment. As a consequence modern agriculture is nowadays confronted with a number of pressing problems. There is an intense debate about the role of agriculture in the diffuse pollution of the environment by the intensive use of agrochemicals and organic manure. These problems reflect negatively on agriculture because they are directly sensed by the society. Farmers themselves are also facing several direct negative consequences of the modern production methods. Agriculture in industrialised societies has to address the degradation of physical soil structure resulting in erosion and soil compaction.

Erosion causes financial damage on the farm through the formation of rills and gullies and the washing away of fertile soil, seeds, manure and fertilizers. The loss of fertile soil by erosion not only has serious effects on crop yields but also negatively affects the soil functions, as it reduces plant rooting depth, removes nutrients and soil organic matter (SOM), reduces infiltration rates and plant available soil water (PASW). Decrease in soil biodiversity is another and very important on-site impact of erosion. Decline in soil biodiversity affects nutrient turnover, decreases aggregate stability, increases crusting, reduces infiltration rates and exacerbates erosion (Pimentel *et al.*, 1995 and Lupwayi *et al.*, 2001). There are also important off-site problems caused by erosion like pollution of drinking water resources, the accelerated silting up of water reservoirs and mud on the roads and in housing properties of densely populated areas. Next to the offsite costs for the society for dredging waterways and cleaning the roads, the muddy floods also result in financial costs for the private households and have an emotional impact on the inhabitants (Pimentel *et al.*, 1995; Verstraeten & Poesen, 1999 and Verstraeten *et al.*, 2003a & 2006). Erosion in Belgium mainly occurs in the loess belt, which stretches from east to west across the central part of the country, i.e. in the south of Flanders, and in the north of Wallonia (Verstraeten & Poesen, 1999; Gobin & Govers, 2003 and Verstraeten *et al.*, 2006). The major erosion problems are found with root crops and maize (Zea mays ssp. Mays L.) (Esteve *et al.*, 2004). The yearly erosion from silt loam soils in the hilly areas in Belgium varies between a few to 100 Mg soil ha^{-1} y^{1} (Verstraeten *et al.*, 2003b).

Next to erosion, soil compaction also seriously threatens the agricultural production in some areas in Europe. Soil compaction essentially reduces the pore space between soil particles and can occur both at the surface and in subsurface soil horizons. The worst effects of surface soil compaction can be rectified relatively easily by soil tillage, root growth and biological activity in general and, hence, it is perceived to be a less serious problem in the medium to long term. On the contrary, once subsoil compaction occurs, it can be extremely difficult and expensive to alleviate. Furthermore, remedial treatments usually need to be repeated. Deep soil compaction decreases the growth of plants by a reduction of the plant rooting depth and PASW (capacity) and often results in a decrease of crop yield (Ide *et al.*, 1984; Ide & Hofman, 1990 and Jones *et al.*, 2004).

As a result of human induced erosion and soil compaction the soil fertility of arable land is diminishing continuously. Erosion and soil compaction strongly depend on management. The main human causes for the degradation of the physical soil structure are the frequent passages of the fields with heavy farming equipment, also under unfavourable circumstances, ploughing and intensive soil tillage in general causing a disruption of aggregates and a decline in SOM, increased farm and field sizes, limited crop rotations and intensification of crops with lower crop cover density. After all, farmer's management decisions are determined by market conditions, technological development and changes in the global economy, particularly the rising relative cost of labour (Esteve *et al.*, 2004; Jones *et al.*, 2004; Bronick & Lal, 2005 and Hamza & Anderson, 2005).

Until the mid 1990's, erosion and its related problems received little attention in the environmental debate. This has changed through the increasing reports on erosion and increased interest in environmental issues in general (Boardman *et al.*, 2003 and Vertraeten et al., 2003a). The last 5 to 10 years RT agriculture is increasing gradually in whole Europe due to an increasing concern and awareness in soil and environmental protection. The recent enthusiasm about RT agriculture can also partly be explained by the progress in agricultural machinery, more specifically the sowing machines. Sowing in crop residues and green manure on the surface indeed demands adapted equipment. Moreover, the economical circumstances force the farmers to reduce the production costs, while the pressure on the environment urges farmers to manage his soil capital better and reduce runoff and erosion (Vandergeten & Roisin, 2004).

Under reduced tillage (RT) agriculture, the soil is not inverted and mixed with the crop residues and this seems to profoundly impact many soil properties particularly in the upper soil layer. In their literature review of runoff and erosion during rainfall simulations and natural rainfall under field conditions in temperate climate, Strauss *et al.* (2003) concluded that RT with direct drilling (RT_{DD}) agriculture mostly decreased runoff and erosion compared to conventional tillage (CT) agriculture. Rainfall simulations in Belgium indicated that RT agriculture generally reduces erosion compared to CT agriculture. However, a large variation in erosion response was observed. The reduction of erosion by RT compared to CT agriculture measured in different experiments varied between 0 to 95% and was on average 35 to 55% (Goyens *et al.*, 2005; Leys *et al.*, 2007 and Vermang *et al.*, 2007). The residues of the crops or green manure at the soil surface indeed intercept the rain and protect the soil against crusting of the soil surface. Moreover, the coverage slows down the runoff flow velocity (Layton *et al.*, 1993; Vandergeten & Roisin, 2004 and Gillijns *et al.*, 2005). Nevertheless, RT fields often have a lower field-saturated hydraulic conductivity K_{fs} compared to CT fields. It has been reported that 3 to 18 years after converting CT to RT_C or RT_{DD} agriculture on loam soils, the K_{fs} was lower or at best comparable under RT_{DD} than RT_C and CT agriculture (Wienhold & Tanaka, 2000; Lipiec *et al.*, 2006 and Singh & Malhi, 2006). On the other hand, Liebig *et al.* (2004) reported that after 15 years of RT_{DD} agriculture on a silt loam soil in the Great Plains, K_{fs} was higher than under CT agriculture. Seeing these contrasting results further research on the effect of a conversion from CT to RT agriculture on K_{fs} seems therefore needed.

Studies by Friedel *et al.* (1996), Hussain *et al.* (1999) and Liebig *et al.* (2004) showed that the aggregate stability of the upper soil layer of fields with a cereal, maize and soybean crop rotation was higher in RT_C or RT_{DD} compared to CT agriculture. The aggregate stability proves to have a negative relationship with runoff and soil erosion (Barthès & Roose, 2002).

Although RT agriculture is growing more important in today's agriculture in Western Europe the crop rotations are somewhat particular because of the large share of root crops. However, no research has been carried out on the effect of RT agriculture on the bulk density (BD) and aggregate stability of soils with crop rotations including beet (*Beta vulgaris* L.) and potatoes (*Solanum tuberosum* L.), with heavy soil disturbance at the formation of the ridges and at harvest, that seem less suitable for RT agriculture.

The objective of this study was to investigate the short- and long- term effects of RT agriculture on aggregate stability, penetration resistance (PR), BD and field-saturated hydraulic conductivity K_{fs} under the specific western European climatic and soil conditions and typical rotations including crops which are often assumed to be less suitable for RT agriculture such as beets and potatoes.

2. Site Description

Belgium has a temperate maritime climate with mild winters and cool summers. Our research sites were situated in the silt loam belt of central Belgium with an average precipitation of 780 mm y^{-1} and an average yearly temperature of 9.8 °C. However, significant deviations from the long term average (30 years) rainfall (690 mm in 2003 and 914 mm in 2004) and temperature (11.1 °C in 2003, 10.7 °C in 2004 and 11.0 °C in 2005) have been observed in recent years (KMI, 2007). The climate parameters are very important because they affect the soil moisture content and by consequence the PR.

In Belgium, CT fields are generally tilled with a mouldboard plough which inverts the soil and buries crop residues to a depth of 25 to 30 cm. Depending on the crop residues, crop rotation and the application of organic manure, ploughing of CT fields is combined with cultivating and/or harrowing. Under RT agriculture, different types of cultivators and soil looseners in combination with harrows are used.

Eighteen fields with a silt loam soil texture in Belgium (Flanders and Wallonia) were selected (D'Haene, 2008). Figure 1 shows the location of the observation sites. The fields included different RT types running for a different number of years (between 2 and 20 years), and were paired to CT fields with comparable soil type and crop rotation (Table 1). Particular care was taken that the texture of the paired RT-CT fields were similar, because of the major influence of texture on C and N dynamics in soil. The tillage operations of 2003, 2004 and spring 2005 are given in Table 2. The period of RT agriculture (in years) is indicated in subscript.

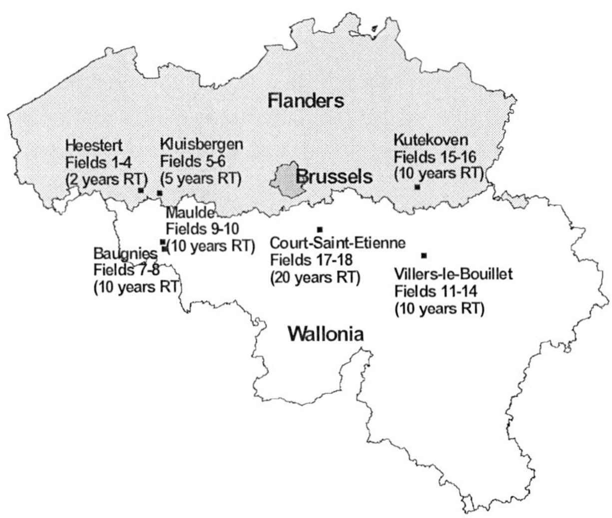

Figure 1. Geographical location of sampling sites

3. MATERIAL AND METHODS

3.1. Selection of Fields

In each field, 3 small plots with a size of 150 m^2 (10 x 15 m) were selected. The plots were separated with 10 m spacing. These 3 plots per field were further considered as replicates for sampling and data analysis. To avoid edge effects, the plots were located more than 20 m from the edges of the field. On sloping fields, the plots of RT and CT were located at the same position on the slope.

3.2. Soil Sampling

The optimal conditions for sampling for measuring the aggregates stability is when clods can be collected and broken apart by hand to produce a maximum of natural aggregates (Le Bissonnais, 1996). Therefore, samples for aggregate stability were taken in 10 replicates per plot from the 0-10 cm layer between 15 June 2005 and 19 July 2005, one month after the sowing period of beets and maize. These samples were then mixed to obtain a representative composite sample. The PR and K_{fs} were in measured in the field at sampling of the aggregate stability. The PR was measured with a hand penetrologger (Eijkelkamp Agrisearch Equipment, the Netherlands), which was plunged into the soil 12 times at random locations per plot (Van der Lecq *et al.*, 2001). Field-saturated hydraulic conductivity K_{fs} was measured in three replicates per plot on fields 11-18. At the same day, undisturbed soil samples (diameter 3.58 cm) in three replicates per plot were taken from the 0-60 cm layer per 5 cm layer from fields 1-6 and 11-18 to determine the BD and soil moisture content. In fields 7-10 soil samples (diameter 1.90 cm) in ten replicates per plot were taken from the 0-60 cm layer per 5 cm to determine the BD and soil moisture content.

Five samples per plot were taken from the 0-10, 10-20, 20-30 and 30-40 cm layers for the determination of soil organic carbon (SOC). In each layer, the samples were bulked per plot into one composite sample, thoroughly mixed and air-dried in the laboratory. Soil texture of each field was determined per layer on a mixed soil sample of the three plots. Fields 1-8 and 17-18 were sampled in December 2004, whereas fields 9-16 were sampled in March 2005.

3.3. Aggregate Stability

The aggregate stability was measured on air dried soil samples using the "dry and wet sieving" method of De Leenheer & De Boodt (1959) (UGent method). The instability index was calculated as the difference of the mean weight diameter (MWD) of the dry sieving minus the MWD of the wet sieving. The inverse of the instability index, i.e. the stability index (SI), was taken as a measure of the stability of the aggregates (De Leenheer & De Boodt, 1959 and De Boodt & De Leenheer, 1967).

Additionally, the aggregate stability was measured according to Le Bissonnais methodology (Le Bissonnais, 1996 and Legout *et al.*, 2005), being fast wetting of the soil simulating a heavy shower "most aggresive" (MWD_{fast}), slow wetting of the soil "least aggresive" (MWD_{slow}) and stirring the soil after prewetting to test the wet mechanical cohesion soil aggregates (MWD_{stir}).

3.4. Penetration Resistance, Bulk Density and Soil Moisture Content

The PR is the resistance of the soil against penetration of a cone and is a measure for the compaction of the soil. The maximum resistance a root can penetrate is 3 MPa (Ide & Hofman, 1990) and this is half of the maximum PR of the used penetrologger. The PR of the studied fields was measured with a hand penetrologger (Eijkelkamp Agrisearch Equipment, the Netherlands), which was plunged into the soil 12 times at random per plot (Van der Lecq *et al.*, 2001). The hand penetrologger (60° angle cone with a base of 1 cm^2) was pushed perfectly straight to 60 cm depth into the soil by applying equal pressure on both grips.

BD and soil moisture content were determined from the oven dry mass (105 °C) of the undisturbed samples and volume of the soil core (Soil Survey Staff, 1992). No shrinkage was observed in any of the cores.

Table 1. Management, period of reduced tillage (RT), percentage clay, loam, sand, of the 0-10 cm and crops of 2003, 2004 and at sampling in 2005 of the selected fields

Field	Management*	Period of RT	Clay	Loam	Sand	Crop rotation **		
		(years)	(g kg^{-1})	(g kg^{-1})	(g kg^{-1})	Crop of 2003	Crop of 2004	Crop at sampling
1	RT_C	2	135	526	339	maize (fodder)	winter wheat/mustard†	sugar beet
2	RT_C	2	124	537	339	maize (fodder)	winter wheat/mustard†	sugar beet
3	RT_C	2	121	599	280	maize (fodder)	winter wheat/mustard†	sugar beet
4	CT	/	127	544	329	maize (fodder)	winter wheat/mustard†	sugar beet
5	RT_C	5	181	516	303	maize (grain)	maize (grain)	maize (grain)
6	CT	/	164	560	276	maize (fodder)	potatoes	maize (fodder)
7	RT_C	5	106	590	304	sugar beet	winter wheat/winter oat†	potatoes
8	CT	/	111	596	293	triticale	winter barley	maize (fodder)
9	RT_C	10	206	709	85	winter wheat	winter barley/mustard†	sugar beet
10	CT	/	139	776	85	maize (fodder)	maize (fodder)	maize (fodder)
11	RT_{DD}	10	198	722	79	winter barley/rape†	sugar beet	winter wheat
12	CT	/	189	754	57	sugar beet	winter wheat/phacelia†	maize (fodder)
13	RT_{DD}	10	167	772	61	winter barley/rape†	potatoes	winter wheat
14	CT	/	162	746	94	winter barley	potatoes	winter wheat
15	RT_{DD}	10	155	717	128	winter wheat	winter barley/rape†	sugar beet
16	CT	/	174	715	111	winter wheat	winter barley/mustard†	sugar beet
17	RT_C	20	147	715	138	winter wheat/mustard†	sugar beet	winter wheat
18	CT	/	160	757	82	winter wheat/mustard†	sugar beet	winter wheat

* RT_C: reduced tillage field with cultivator or soil loosener; RT_{DD}: reduced tillage field by direct drilling, CT: conventional tillage field

** maize: *Zea mays* ssp. *Mays* L.; winter wheat: *Triticum aestivum* L.; mustard: *Sinapis alba* L.; sugar beet: *Beta vulgaris* L.; potatoes: *Solanum tuberosum* L.; winter oat: *Avena sativa* L.; triticale: *X Triticosecale*; winter barley: *Hordeum vulgare* L.; rape: *Brassica rapa* L.; phacelia: *Phacelia* L.

†: green manure

Table 2. Type and depth of combined tillage operations of 2003, 2004 and spring 2005

Field*		Tillage**	Depth (cm)	Tillage	Depth (cm)	Tillage	Depth (cm)	Tillage	Depth (cm)	Time***		
1	RT_{C_2}	cultivator	10-15	tine harrow	5	tine harrow	5			2003_S	2003_A	2005_S
		cultivator	10-15	tine harrow	5					$2004_A^\dagger$		
2	RT_{C_2}	cultivator	10-15	soil loosener	15-20	tine harrow	5	tine harrow	5	2003_S	2003_A	2005_S
		cultivator	10-15	tine harrow	5					$2004_A^\dagger$		
3	RT_{C_2}	cultivator	10-15	soil loosener	15-20	tine harrow	5	tine harrow	5	2003_S	2003_A	2005_S
		cultivator	10-15	tine harrow	5					$2004_A^\dagger$		
4	CT	cultivator	10-15	plough	25-30	tine harrow	5	tine harrow	5	2003_S	2003_A	2005_S
		cultivator	10-15	tine harrow	5					$2004_A^\dagger$		
5	RT_{C_5}	cultivator	10	soil loosener	35	cultivator	10	rotary harrow	5	2003_S	2004_S	
		cultivator	10	soil loosener	25	rotary harrow	5			2005_S		
6	CT	cultivator	10	plough	25-30	cultivator	10	rotary harrow	5	2003_S	2004_S	2005_S
7	RT_{C_5}	cultivator	5	cultivator	12	rotary harrow	5			2003_S	2005_S	
		cultivator	15	cultivator	15					2003_A		
		soil loosener	25	rotary harrow	5					$2004_A^\dagger$		
8	CT	cultivator	10	plough	30	cultivator	10	rotary harrow	5	2003_A	2004_A	
9	RT_{C_10}	soil loosener	20	rotary harrow	5					2003_A		
		cultivator	10	rotary harrow	5					2004_A	2005_S	
10	CT	cultivator	10	plough	25-30	cultivator	10	rotary harrow	5	2003_S	2004_S	2005_S
11	RT_{DD_10}	rotary harrow	5							$2003_A^\dagger$		
12	CT	plough	25-30	rotary harrow	5					2003_S	2003_A	
		soil loosener	25-30	rotary harrow	5					$2004_A^\dagger$		
		rotary harrow	5							2005_S		
13	RT_{DD_10}	soil loosener	25	rotary harrow	5					$2003_A^\dagger$		
14	CT	cultivator	10	plough	25-30	cultivator	10	rotary harrow	5	2004_S		
		plough	25-30	cultivator	10	rotary harrow	5			2004_A		
15	RT_{DD_10}	rotary harrow	5							2003_A	$2004_A^\dagger$	
16	CT	plough	20-25	tine harrow	5					2003_A		

Table 2.(Continued)

Fie ld*	Tillage **	Depth (cm)	Tillage	Depth (cm)	Tillage	Depth (cm)	Tillag e	Depth (cm)	Tim e***			
		cultivator	10							2004_A		
		plough	20-25	cultivator	10	tine harrow	5			2005_W		
17	RT_{C_20}	cultivator	10	cultivator	10	soil loosener	25	rotary harrow		$2003_A^{\dagger}$		
		rotary harrow	5							2004_S		
		soil loosener	25	rotary harrow	5					2004_A		
18	CT	cultivator	10	cultivator	10					$2003_A^{\dagger}$		
		plough	25	rotary harrow	5	rotary harrow	5			2004_S		
		soil loosener	25	rotary harrow	5					2004_A		

* RT_C: reduced tillage field with cultivator or soil loosener with number of years in subscript; RT_{DD}: reduced tillage field by direct drilling with number of years in subscript; CT: conventional tillage field

**: the successive tillage operations are given per row

***: $_S$: spring (March – May) tillage operation; $_A$: autumn (August – November) tillage operation, $_W$: winter (December-February) tillage operation, †: tillage operation before green manure

3.5. Field-Saturated Hydraulic Conductivity

The Guelph pressure infiltrometer with 9.7 cm inner diameter (Soil Moisture Equipment Corp, USA) has been used to measure field - satured hydraulic conductivity (K_{fs}) using the single head method (Reynolds & Elrick, 2002). Using the single head method with the soil-structure parameter α^* (see e.g. Reynolds & Elrick, 2002) taken from Elrick *et al.* (1989) to calculate K_{fs} is often sufficient for practical applications. The method involves measuring the steady-state rate of water infiltration in which a constant depth (head) of water is maintained. A "bulb" of saturated soil with specific dimensions under the ring is rather quickly established by the infiltrometer. This bulb is very stable and its shape depends on the type of soil, the radius of the ring and the head of water in the ring (Reynolds & Elrick, 2002). Using the equation of Reynolds & Elrick (1990) which is based on a solution for three-dimensional flow, K_{fs} can then be calculated from the steady-state infiltration rate (Phylip, 1985; Reynolds & Elrick, 1991; Soil Moisture Equipment Corp, 1992 and Mathieu & Pieltain, 1998).

3.6. Soil Organic Carbon and Soil Texture

The SOC content was analyzed according to the method of Walkley & Black (1934). Soil texture of each field was determined per layer using the combined sieve and pipette method (De Leenheer, 1959).

4. Statistical Analysis

The obtained results were first analyzed per site, to minimize variability and increase the comparability of the aggregate stability, BD, PR and K_{fs}. The homogeneity of variances was tested with the Levene's test ($P = 0.05$). A *t*-Test was used to find significant differences for locations with only 2 fields. One way analysis of variances (ANOVA) with field as factor/post hoc Duncan test and Welch/post hoc Games-Howell test were used to determine significant differences for the locations with more than 2 fields for homogeneous and heterogeneous variances, respectively. Statistical correlation of the data was done using linear regression analysis. A correlation analysis was performed using a Pearson's correlation matrix in SPSS (version 12.0, SPSS Inc., USA).

5. Results

5.1. Relation between Aggregate Stability and Soil Tillage

The aggregate stability measured with the different procedures showed the same trend in fields 1 to 4 in Heestert, but the results had a different significance (Figure 2 and 3). The aggregate stability was the highest for all procedures in field 2 (RT_{C_2}) and the lowest in field 4 (CT). The biggest difference between the fields was found in MWD_{slow}. Since fields 1-3 were only for 2 years under RT, no big differences were expected. After 5 years RT_C

agriculture significant differences were found for SI and MWD_{slow} and MWD_{stir} (RT_{C_5} on field 5 compared to CT field 6, RT_{C_5} on field 7 compared to CT field 8). There was no obvious trend in MWD_{fast} after 5 years RT_C agriculture. After 10 years RT_C agriculture, the SI and MWD in the three methods was significantly higher on RT_{C_10} field 9 compared to CT field 10. When comparing fields 11 to 14 (RT_{DD_10} vs. CT), CT field 14 had the lowest SI. RT_{DD_10} fields 11 and 13 had significantly higher MWD_{slow} and MWD_{stir} values than CT fields 12 and 14, respectively. On the other hand, MWD_{stir} was significantly lower when comparing RT_{DD_10} field 15 with CT field 16. In general we found that aggregate stability after 10 years RT followed the order RT_{DD_10} > RT_{C_10} > CT fields. After 20 years RT_C agriculture, there was an insignificant increase in SI in RT_{C_20} field 17 compared to CT field 18. The MWD_{fast}, MWD_{slow} and MWD_{stir} were significantly higher in RT_{C_20} field 17 than CT field 18.

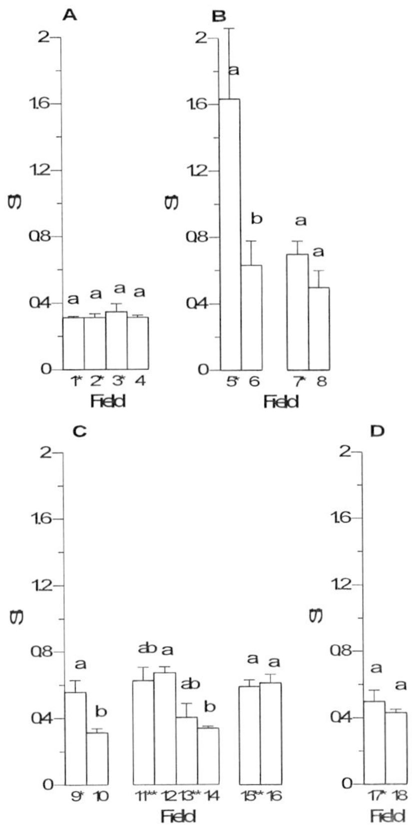

*: reduced tillage with cultivator or soil loosener (RT_C)

**: reduced tillage by direct drilling (RT_{DD})

A: 2 years RT_C
B: 5 years RT_C
C: 10 years RT_C or RT_{DD}
D: 20 years RT_C

same letter indicates no significant differences between tillage treatments at P = 0.05 per location (one way ANOVA/Duncan post hoc test, Welch/Games -Howell post hoc test or *t*-Test)

Figure 2. Stability index (SI) (line = standard deviation) determined by the 'dry and wet sieving' method of De Leenheer & De Boodt (1959) of the 0-10 cm depth layer of the 18 selected fields

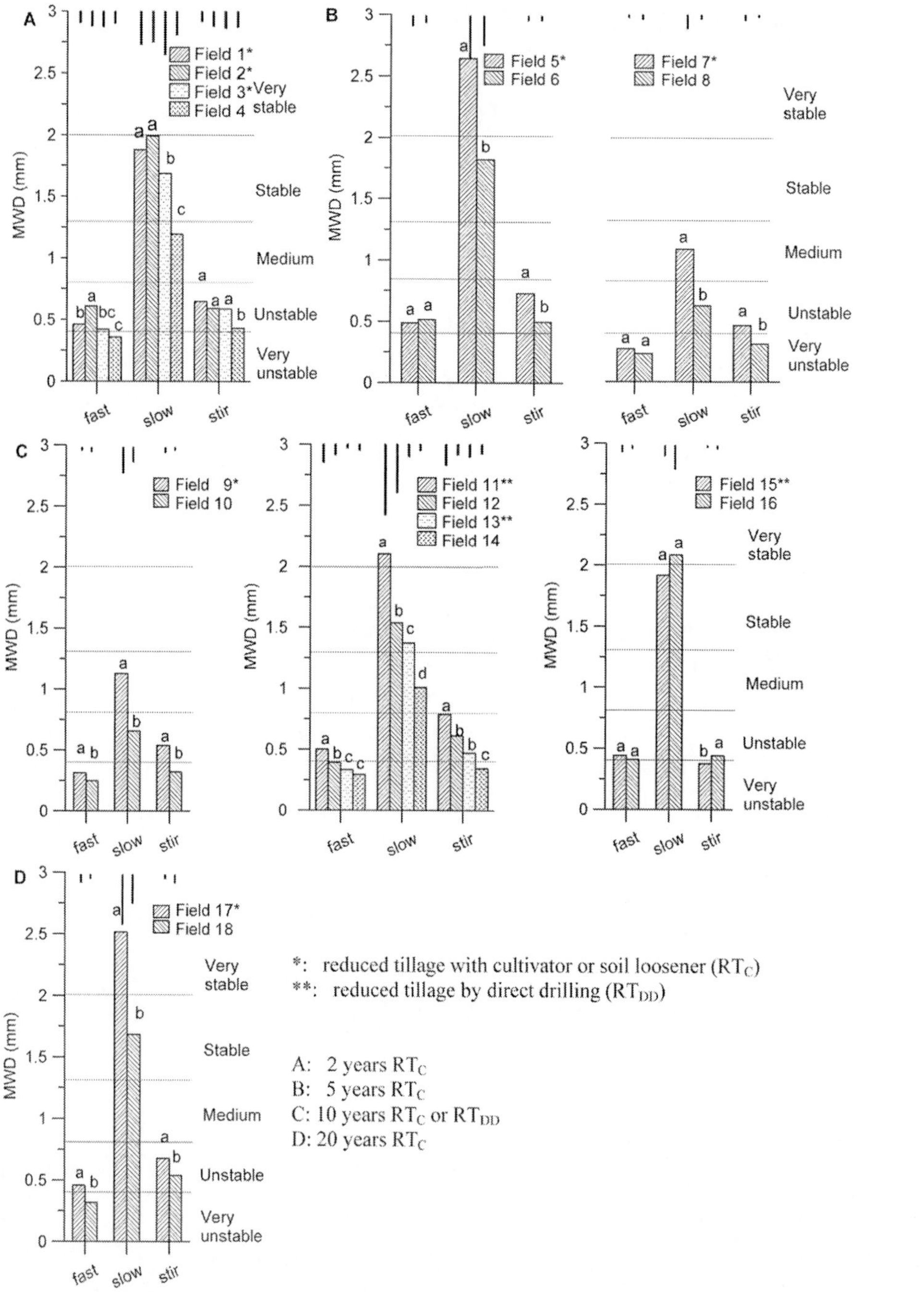

same letter per location indicates no significant differences between tillage treatments at P = 0.05 (one way ANOVA/Duncan post hoc test, Welch/Games-Howell post hoc test or *t*-Test) (vertical lines = standard deviation)

Figure 3. Mean weight diameter (MWD) (mm) measured by the 3 methods of Le Bissonnais (1996) of the 0-10 cm depth layer of the 18 selected fields

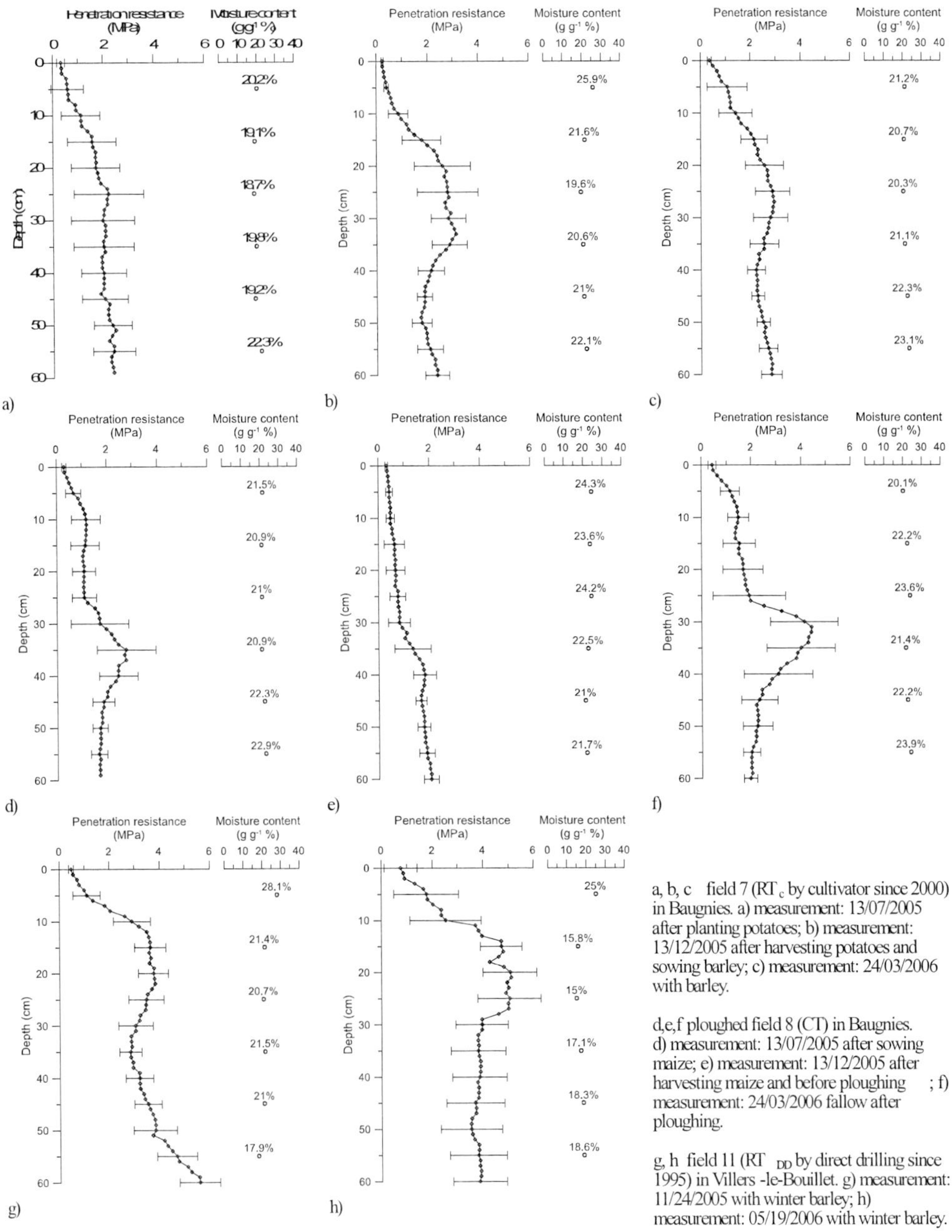

a, b, c field 7 (RT_c by cultivator since 2000) in Baugnies. a) measurement: 13/07/2005 after planting potatoes; b) measurement: 13/12/2005 after harvesting potatoes and sowing barley; c) measurement: 24/03/2006 with barley.

d,e,f ploughed field 8 (CT) in Baugnies. d) measurement: 13/07/2005 after sowing maize; e) measurement: 13/12/2005 after harvesting maize and before ploughing ; f) measurement: 24/03/2006 fallow after ploughing.

g, h field 11 (RT_{DD} by direct drilling since 1995) in Villers -le-Bouillet. g) measurement: 11/24/2005 with winter barley; h) measurement: 05/19/2006 with winter barley.

Figure 4. Penetration resistance (MPa) and soil moisture content (g g^{-1} %) of fields 7, 8 and 11

5.2. Penetration Resistance and Bulk Density under Different Tillage Systems

The PR has a high spatial variability. The high heterogeneity makes it difficult to compare the PR of the different tillage systems and to draw conclusions. The PR also varies with time. Shortly after field work (June – July 2005) a low PR was measured in the RT_C and

CT fields. During the growing season or winter period the PR of the RT_C but especially of the CT fields was increased (e.g. RT_{C_5} 7 and CT 8 fields in Figure 4). The PR of fields 5, 6, 11, 13, 15 and 16 was too high to measure in June – July 2005 with the hand penetrologger due to the low soil moisture conditions.

The PR of the 0-5 cm depth layer was comparable for all fields. The PR of the 5-30 cm depth layer was higher in the soil profile under RT_{DD} compared to RT_C and CT fields (e.g. RT_{DD_10} field 11 compared to RT_{C_5} 7 and CT 8 fields in Figure 4). We can see a plough pan at 30 to 40 cm depth in field 8 (CT) (Figure 4). In CT fields 8, 10, 12, 14, 16 and 18 also an obvious plough pan was measured. The plough pan could however not be measured when the soil moisture conditions were too high (e.g. the plough pan was not measured on 13/12/2005 in CT field 8 (Figure 4e)).

BD varied with depth in the same trend as the PR (Figure 5 and Table 3). The BD of RT_{DD} was higher than the RT_C and CT fields in the 0-30 cm layer. The BD of CT fields was higher in the 30-40 cm layer i.e. CT field 12 compared to RT_{DD_10} field 11 if a plough pan was present. Deeper in the soil profile no difference in BD between the different tillage systems could be found.

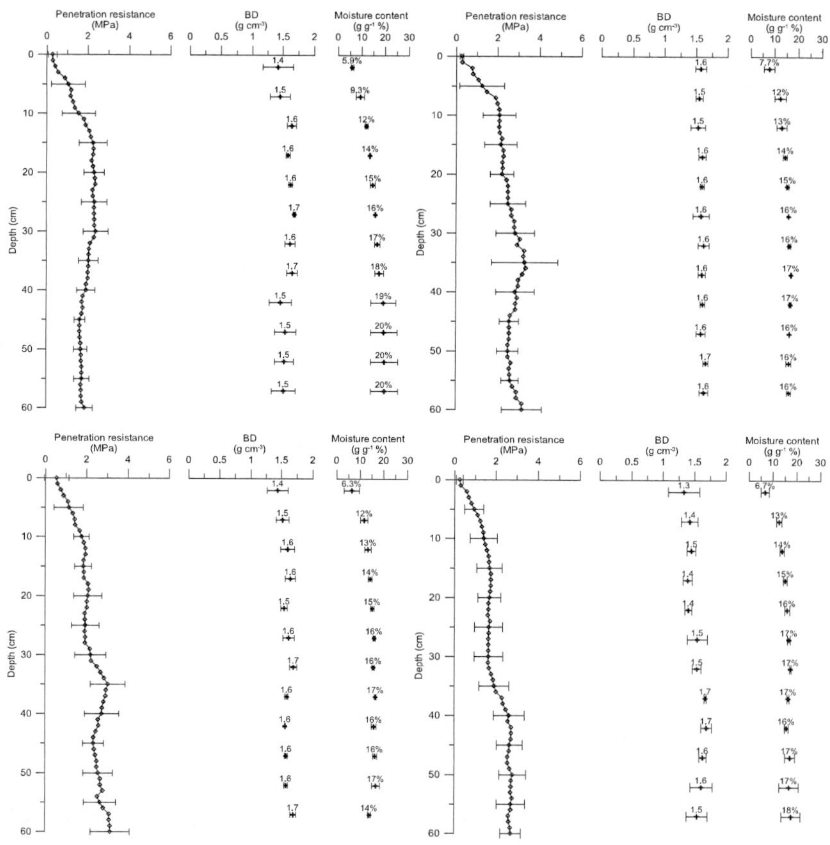

Figure 5. Penetration resistance (MPa), bulk density (BD) (g cm^{-3}) and soil moisture content (g g^{-1} %) of fields 1-3 (reduced tillage with cultivator or soil loosener since 2003) and ploughed field 4 measured with a hand penetrologger in Heestert (22/06/2005)

Table 3. Bulk density of the 0-60 cm layer (g cm^{-3}) with standard deviation between brackets) of fields 11-18 measured in June-July 2005

	11		12		13		14		15		16		17		18	
	RT_{DD_10}		CT		RT_{DD_10}		CT		RT_{DD_10}		CT		RT_{C_20}		CT	
0-5	1.24	(0.18)	1.05	(0.20)	1.43	(0.13)	1.27	(0.08)	1.26	(0.11)	1.27	(0.03)	1.33	(0.10)	1.35	(0.06)
5-10	1.53	(0.19)	1.47	(0.21)	1.47	(0.21)	1.42	(0.02)	1.52	(0.12)	1.66	(0.05)	1.48	(0.07)	1.51	(0.08)
10-15	1.59	(0.15)	1.56	(0.12)	1.71	(0.16)	1.29	(0.13)	1.55	(0.05)	1.61	(0.11)	1.43	(0.06)	1.65	(0.14)
15-20	1.63	(0.19)	1.54	(0.14)	1.63	(0.15)	1.25	(0.19)	1.57	(0.13)	1.67	(0.19)	1.62	(0.17)	1.59	(0.08)
20-25	1.61	(0.14)	1.52	(0.12)	1.75	(0.02)	1.29	(0.14)	1.65	(0.09)	1.51	(0.06)	1.52	(0.19)	1.62	(0.10)
25-30	1.53	(0.15)	1.55	(0.09)	1.53	(0.12)	1.52	(0.11)	1.67	(0.08)	1.62	(0.12)	1.62	(0.11)	1.62	(0.10)
30-35	1.56	(0.17)	1.65	(0.06)	1.59	(0.02)	1.62	(0.15)	1.59	(0.09)	1.55	(0.09)	1.64	(0.08)	1.54	(0.12)
35-40	1.55	(0.04)	1.64	(0.03)	1.52	(0.14)	1.62	(0.06)	1.63	(0.10)	1.59	(0.17)	1.68	(0.06)	1.68	(0.05)
40-45	1.68	(0.12)	1.62	(0.06)	1.50	(0.08)	1.62	(0.04)	1.64	(0.11)	1.71	(0.08)	1.66	(0.07)	1.47	(0.10)
45-50	1.51	(0.03)	1.60	(0.11)	1.52	(0.12)	1.48	(0.02)	1.48	(0.04)	1.62	(0.07)	1.52	(0.07)	1.49	(0.10)
50-55	1.65	(0.15)	1.61	(0.11)	1.58	(0.12)	1.54	(0.02)	1.56	(0.15)	1.66	(0.01)	1.65	(0.12)	1.67	(0.10)
55-60	1.53	(0.09)	1.60	(0.09)	1.65	(0.06)	1.49	(0.10)	1.56	(0.09)	1.51	(0.05)	1.63	(0.05)	1.63	(0.13)

Table 4. Field-saturated hydraulic conductivity (K_{fs}) and soil moisture content of the 0-5 cm depth layer (with standard deviation between brackets) in June – July 2005 in fields 11-18

field		K_{fs} (cm h^{-1})		soil moisture content (%)	
11	$RT_{DD\ 10}$	0.7	(0.3)	8.8	(0.3)
12	CT	0.6	(0.4)	11.2	(0.8)
13	$RT_{DD\ 10}$	0.8	(0.7)	10.2	(0.5)
14	CT	1.2	(0.3)	5.6	(0.9)
15	$RT_{DD\ 10}$	1.1	(0.4)	9.7	(0.4)
16	CT	0.6	(0.3)	11.8	(0.3)
17	$RT_{C\ 20}$	5.8	(8.8)	15.7	(1.9)
18	CT	0.4	(0.1)	16.6	(0.5)

- : no results

5.3. Field-Saturated Hydraulic Conductivity and Tillage Systems

Per location there were no significant differences in K_{fs} between the fields 11–18 (Table 4), but the high variability in K_{fs} between the plots of some fields complicated the comparison. The general results indicate that in RT fields the K_{fs} were higher than under CT.

6. DISCUSSION

6.1. Aggregate Stability under Different Tillage Systems

There is an enormous variety in the methods for measuring the aggregate stability, which complicates a comparison of the results. The retention of crop residues at the soil surface exerts a positive influence on the formation of aggregates even in the short term (Hermawam & Bomke, 1997; Martens, 2000 and Coppens *et al.*, 2006). Bossuyt *et al.* (2001) measured a higher aggregate stability after addition of crop residues, which was higher for the crop residues with a higher C:N ratio. Changing from CT agriculture to RT agriculture under temperate climate conditions mostly resulted in a (significant) higher aggregate stability in the upper layers of RT fields (Friedel *et al.*, 1996; Hussain *et al.*, 1999; Diaz-Zorita *et al.*, 2002 and Liebig *et al.*, 2004).

The results of our study confirm that the reduction of tillage intensity and the retention of crop residues at the soil surface of RT fields result in a higher aggregate stability of the RT compared to CT fields despite the frequent disturbance on the occasion of harvest of root crops might every two to three years (Figures 2 and 3).

In our research we observed an increased aggregate stability with the period of RT_C, as was found by Alakukku (1998) and Alakukku *et al.* (2003). It was expected that RT_{DD_10} fields 11 and 13 would have more or less the same aggregate stability because the tillage was the same for 10 years (6 years reduced tillage followed by 4 years direct drilling). However, the agregate stability of RT_{DD_10} field 13 was lower than RT_{DD_10} field 11. This could probably be explained by the potatoes grown in RT_{DD_10} field 13 causing soil disturbance at the formation of the ridges and at harvesting, while the beets of RT_{DD_10} field 11 only caused soil disturbance at harvesting. Kuttekoven (RT_{DD_10} field 15 and CT field 16) was the only site where the aggregate stability was lower under RT_{DD} than under CT (but not statistically significant).

The results obtained with the "dry and wet sieving" method of De Leenheer & De Boodt (1959) (UGent method) and MWD_{slow} measured according to Le Bissonnais (1996) methodology yielded a better discrimination between the different tillage practices. This is probably linked to the fact that these procedures approach the processes in the field. The "dry and wet sieving" method imitates the drop collision to the different size aggregates, and MWD_{slow} simulates a slow wetting of the aggregates with time. The procedures for MWD_{fast} and MWD_{stir} were too aggressive with the aggregates, resulting in a low aggregate stability in most cases.

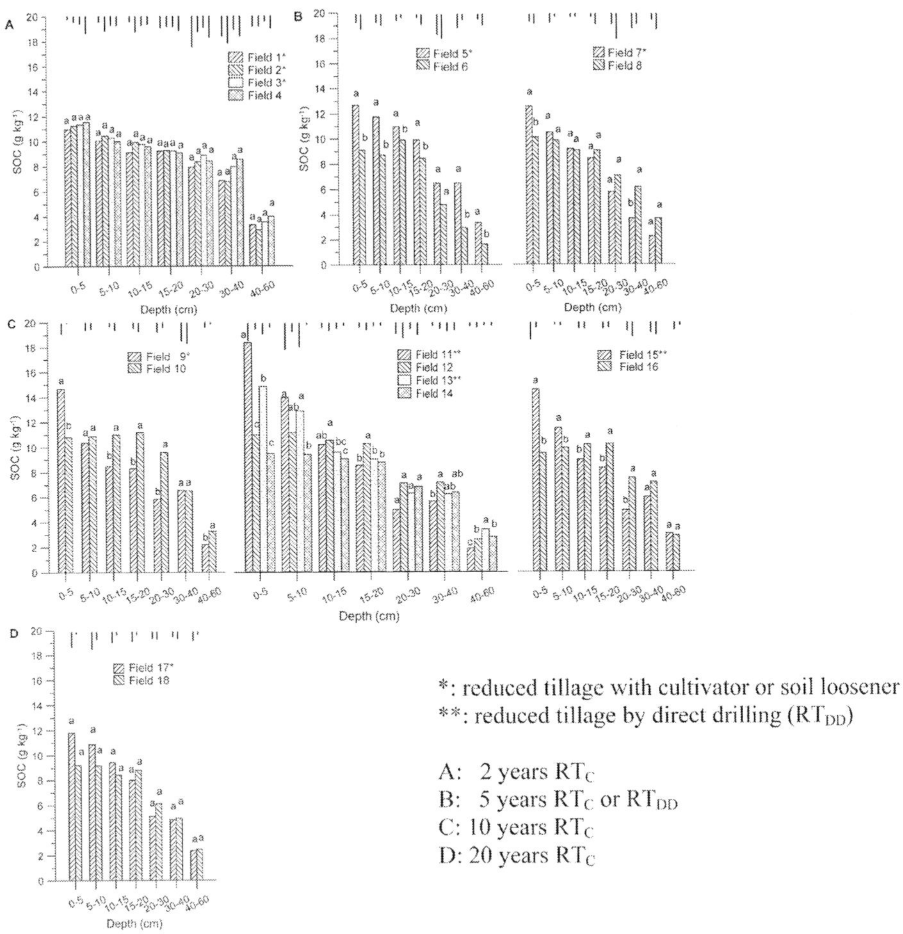

same letters indicate no significant differences between tillage treatments per location per depth layer (P = 0.05) (one way ANOVA/Duncan post hoc test or *t*-Test)

Figure 6. Soil organic carbon (SOC) (g kg^{-1}) (vertical lines = standard deviation) of the 0-5, 5-10, 1-15, 15-20, 20-30, 30-40 and 40-60 cm depth layers of the 18 selected fields (D'Haene *et al.*, 2009)

Table 5. Pearson correlation between mean weight diameter after fast wetting (MWD_{fast}), MWD after slow wetting (MWD_{slow}), MWD after stirring (MWD_{stir}), stability index (SI), bulk density (BD), clay and soil organic carbon (SOC)

		MWD_{slow} (mm)	MWD_{stir} (mm)	SI	BD (g cm^{-3})	Clay (g kg^{-1})	SOC (g kg^{-1})
All fields	MWD_{fast} (mm)	0.83**	0.67**	0.29	0.32	-0.09	0.16
	MWD_{slow} (mm)		0.75**	0.60**	0.16	-0.06	0.21
	MWD_{stir} (mm)			0.47*	0.07	0.03	0.41
	SI				-0.05	0.38	0.22
RT	MWD_{fast} (mm)	0.76*	0.54	0.11	0.41	-0.37	-0.10
	MWD_{slow} (mm)		0.65*	0.58	0.13	-0.25	-0.01
	MWD_{stir} (mm)			0.39	0.05	-0.03	0.11
	SI				-0.03	0.36	0.16
CT	MWD_{fast} (mm)	0.83**	0.65*	0.74*	-0.14	0.37	-0.34
	MWD_{slow} (mm)		0.73*	0.78**	-0.27	0.20	-0.41
	MWD_{stir} (mm)			0.75*	-0.74*	0.15	0.00
	SI				-0.56	0.58	-0.22

**: significant at P = 0.01; *: significant at P = 0.05
RT: reduced tillage, CT: conventional tillage

Under RT tillage the soil is not mixed, and the SOC in the upper layer was higher than in the CT fields (D'Haene *et al.*, 2009) (Figure 6). It was expected that SOC was correlated with aggregate stability since both parameters are related to tillage system. The MWD_{slow} and MWD_{stir} of silt loam soils in Brittany showed a clear linear correlation with SOC content (up to maximum 15 g kg^{-1}) and clay content (up to a maximum of 300 g kg^{-1}). The high differences in clay content of the monoculture maize fields of Le Bissonnais *et al.* (2002) resulted in a high difference in SOC content in the upper layers and a significant correlation between MWD and the soil parameters. Our fields with a high soil disturbance at the harvest of beet or potatoes resulted in a low increase in SOC content, MWD and SI. The lack of correlation between the MWD or SI with texture and SOC content in this research (Table 5) is obviously related to the small range in soil texture and SOC content (D'Haene *et al.*, 2008).

6.2. Penetration Resistance and Bulk Density under Different Tillage Systems

The PR varies during the year and depends on the time since the tillage operations and soil moisture content. The heterogeneity of PR is even higher under RT than CT agriculture. Soil loosening instead of ploughing indeed was found to result in a less physical homogeneous soil, because lower soil moisture conditions and BD are created on the locations where the tines have worked the soil compared to the unworked part of the soil (Perfect & Caron, 2002; Duiker & Beegle, 2006 and Roisin, 2007). The high heterogeneity makes it difficult to compare the PR of the different tillage systems and draw conclusions.

A general conclusion for the differences in PR under RT_C compared to CT agriculture is difficult because the quality of loosening the soil depends on three major factors: the moisture content of the soil, the initial soil structure and the used cultivator or soil loosener (Vandergeten & Roisin, 2004). If the working depth under RT_C agriculture was lower compared to CT agriculture, the PR was higher in the 20-30 cm depth layer of RT_C fields e.g. the higher PR in the 15-20 cm depth layer of RT_C field 1 compared to fields 2-4. Other researchers also found that the changed depth and intensity of the field work in RT_C compared to CT agriculture influenced the PR. After 8 years RT_C agriculture a higher PR was measured in the 10-25 cm in a silt clay loam under RT_C compared to CT agriculture in Sweden. In the 0-10 and 25-50 cm depth layers, there were no differences in PR. The depth of disturbance under RT_C agriculture was 12 cm instead of 25 cm under CT agriculture (Stenberg *et al.*, 2000). After 12 years of RT_C agriculture on a sandy loam soil in Norway and 15 years RT_C agriculture on a silt loam soil in Sweden a higher PR was measured compared to CT agriculture due to the lower depth of disturbance. This was correlated with a higher BD and lower % SOC (Etana *et al.*, 1999 and Riley *et al.*, 2005).

If the farmers cannot harvest the beets and potatoes under optimal soil moisture conditions the soil structure is reduced for several years. If the tillage operation and harvest can occur under optimal soil moisture conditions, RT_C agriculture including root and tuber crops (and green manure) can maintain but not improve the soil structure. As a consequence it is necessary that the soil structure is optimal before changing the management to RT agriculture (Vandergeten & Roisin, 2004 and D'Haene, 2008).

The PR of the RT_{DD} sometimes exceeded the maximum PR of roots (= 3MPa) (Ide & Hofman, 1990).

The soil moisture content ideal to measure the PR is between 20 % and 25%. At sampling in June – July 2005, we could not do measurements from e.g. RT_{DD} fields, because of the dryness of the soils. This is an inconvenience of the hand penetrologger because it is very difficult to press the rod into the soil with an homogeneous pressure and constant speed in dry condition.

PR varied with depth in the same trend as the BD (Figure 5). The same trend in differences in BD was found in this research as by other researchers. Under a temperate climate most researchers report comparable or higher BD in the 0-5 cm layer of short term (≤ 11 years) RT compared to CT fields with cereals, maize and soybean (*Glycine max*) (Angers *et al.*, 1997; Yang & Wander, 1999; Kay & Vanden Bygaart, 2002; Puget & Lal, 2005 and Al-Kaisi *et al.*, 2005), while under long-term RT fields the BD is comparable or lower than on CT fields (Tebrügge & Düring, 1999; Deen & Kataki, 2003 and Dolan *et al.*, 2006). In the 5-20 cm soil layer, mostly no differences in BD were measured under short-term RT compared to CT fields, while for long-term RT fields BD decreased in following order: $RT_{DD} \geq RT_C \geq CT$. Deeper in the soil profile, BD was comparable for RT and CT fields (Angers *et al.*, 1997; Tebrügge & Düring, 1999; Yang & Wander, 1999; Kay & Vanden Bygaart, 2002; Deen & Kataki, 2003; Al-Kaisi *et al.*, 2005; Puget & Lal, 2005 and Dolan *et al.*, 2006). PR varied with depth in the opposite trend as the SOC.

Since roots provide the crops with water and nutrients, a good rooting system is necessary for plant nutrition. For an optimal root growth a homogeneous and loose soil is needed. Compacted zones and cavities in the soil cause a branching or deformation of the roots and as a consequence decrease the crop yield (Pardo *et al.*, 2000 and Vandergeten & Roisin, 2004).

6.3. Correlation between Penetration Resistance and Crop Growth

Ploughing results in a higher soil temperature and dries out the soil under CT compared to RT agriculture. As a consequence CT fields can be sown earlier and the early season growth of RT fields is delayed compared to CT fields (Drury *et al.*, 1999; Balesdent *et al.*, 2000; Larney *et al.*, 2003 and Six *et al.*, 2004).

The only difference between the RT and CT fields of field experiments is the type of tillage. The timing of soil cultivation and harvest and the choice of the crop variety are because of practical considerations the same which facilitates the comparison of soil properties (Powlson, 2007). Due to the delayed growth of crop yields of RT agriculture an underestimation can be observed in field experiments (D'Haene, 2008).

The inferiority in the crop growth under RT_C compared to CT agriculture is decreased rather fast. The crop yields of beets, potatoes and cereal crops were comparable (85-115%) for RT_C and CT agriculture if the soil structure is optimal and tillage operations were done correctly and non superficial (0-25/30 cm depth layer) (Ekeberg & Riley 1996; Riley, 1998; Mehdi *et al.*, 1999; El Titi, 2001; Debout, 2004; Rücknagel *et al*, 2004; Vandergeten & Roisin, 2004; Dam *et al.*, 2005 and Riley *et al.*, 2005). In dry and wet years slightly higher and lower crop yields could sometimes be observed, respectively (Debout, 2004). If the tillage operations were not done under optimal soil conditions, the crops, especially root crops, could have large amounts of branched roots which resulted in a higher amount of soil tare and resulted in lower yields, especially in dry years (Vandergeten & Roisin, 2004). Undeep loosening of the soil (0-15 cm depth layer) could result in a yield decrease under RT_C compared to CT agriculture (Debout, 2004).

The crop yields under RT_{DD} agriculture were comparable or lower than CT agriculture. The reductions in crop yields under RT_{DD} agriculture were probably correlated with the higher PR in the soil profile (Debout, 2004).

6.4. Field-Saturated Hydraulic Conductivity under Different Tillage Systems

The K_{fs} (field - satured hydraulic conductivity) is an indicator of the state of the soil structure. Our results confirm the high variability of K_{fs} (Reynolds *et al.*, 2002). Next to the dryness of the soils, other factors must have an impact on the K_{fs}. According to Matula (2003) the values of K_{fs} in CT fields fluctuate due to the quality of ploughing, and normally the heterogeneity of the top layer in CT fields is important.

We observed a higher K_{fs} under RT compared to CT agriculture in June-July 2005 after autumn tillage operations in RT fields 15 and 17 in 2004 and no tillage operations in the previous 12 months in RT fields 11 and 13. The higher K_{fs} under RT compared to CT agriculture can possibly be explained by the higher aggregate stability, the fact that the channels made by earthworms and roots were less (RT_C) or not (RT_{DD}) destroyed compared to CT fields and vertical cracks from loosening the soil with a cultivator or soil loosener. The very high variability of K_{fs} of RT_{C_20} field 17 was possibly caused by the presence of these natural or manmade channels. The tines of the machines used to loosen the soil indeed make vertical cracks in the soil which transport the water fast to the deeper depth layers. Liebig *et al.* (2004) also measured a higher K_{fs} with an infiltrometer in RT compared to CT fields in the spring after autumn tillage. However, often a lower saturated K_{fs} is measured in RT compared to CT fields 2-3 months after the tillage operations (Wienhold & Tanaka, 2000; Lipiec *et al.*, 2006 and Singh & Malhi, 2006). This suggests that 2-3 months after the tillage operations, the K_{fs} of RT fields is lower than the CT field, while after a longer period (> 6 months) of stabilisation the opposite can be found (D'Haene *et al.*, 2008). However, under long-term RT_{DD} agriculture a compacted crust can be formed at the surface which can decrease K_{fs} and increase erosion.

7. Conclusions

The moment of sampling is very important when measuring soil structural properties. The best procedure to analyze the aggregate stability is the "dry and wet sieving" method of De Leenheer & De Boodt (1959) (UGent method) and MWD_{slow} measured according to Le Bissonnais (1996) methodology. Aggregate stability was higher in RT than in CT, and increases with the period of RT. In general the K_{fs} were higher in RT than in CT, but in RT_{DD} field 13 we found a decreased K_{fs} which was probably caused by a compacted superficial layer.

We suggest the application of RT_C practices in the Western European fields, because it maintains or increases aggregate stability and K_{fs}. This research indicated that short term RT_{DD} agriculture under crop rotations with root crops often result in a lower crop yield, which is probably correlated with the higher PR in the 10-30 cm depth layer. Therefore, the potential of RT_{DD} agriculture in Western Europe is probably limited.

Acknowledgment

This research was financed by the Flemish Institute for the Promotion of Innovation by Science and Technology in Flanders (IWT). We thank our partners from the Interreg project MESAM (Erosion control measures and sensibilization of farmers for the protection of the environment) and from the Walloon Agricultural Research Centre for their competent assistance in selecting the fields and Olle Victoor, Luc Deboosere and Wouter Schiettecatte for their skilful assistance in the field and laboratory.

References

Alakukku, L. (1998). Properties of compacted fine textured soils as afected by crop rotation and reduced tillage. *Soil Till. Res.*, *47*, 83-89.

Alakukku, L., Weisskopf, P., Chamen, W. C. T., Tijink, F. G. J., van der Linden, J. P., Pires, S., Sommerf, C. & Spoor, G. (2003). Prevention strategies for field traffic -induced subsoil compaction: a review Part 1. Machine/soil interactions. *Soil Till. Res.*, *73*, 145-160.

Al-Kaisi, M. M., Yin, X. & Licht, M. A., (2005). Soil carbon and nitrogen changes as influenced by tillage and cropping systems in some Iowa soils. *Agr. Ecosyst. Environ*, *105*, 635-647.

Angers, D. A., Bolinder, M. A., Carter, M. R., Gregorich, E. G., Drury, C. F., Liang, B. C., Voroney, R. P., Simard, R. R., Donald, R. G., Beyaert, R. P. & Martel, J. (1997). Impact of tillage practices on organic carbon and nitrogen storage in cool, humid soils of eastern Canada. *Soil Till. Res.*, *41*, 191-201.

Barthès, B. & Roose, E., (2002). Aggregate stability as an indicator of soil susceptibility to runoff and erosion: validation at several levels. *Catena, 47*, 133-149.

Balesdent, J., Chenu, C. & Balabane, M. (2000). Relationship of soil organic matter dynamics to physical protection and tillage. *Soil Till. Res.*, *53*, 215-230.

Boardman, J., Poesen, J. & Evans, R. (2003). Socio-economic factors in soil erosion and conservation. *Environ. Sci. Policy*, *6*, 1-6.

Bossuyt, H., Denef, K., Six, J., Frey, S. D., Merckx, R. & Paustian, K. (2001). Influence of microbial populations and residue quality on aggregate stability. *Appl. Soil Ecol. 16*, 195-208.

Bronick, C. J. & Lal, R. (2005). Soil structure and management: a review. *Geoderma*, *124*, 3-22.

Coppens, F., Merckx, R. & Recous, S. (2006). Impact of crop residue location on carbon and nitrogen distribution in soi land water-stable aggregates. *Eur. J. Soil Sci.*, *57*, 570-582.

Dam, R. F., Mehdi, B. B., Burgess, M. S. E., Madramootoo, C. A., Mehuys, G. R. & Callum, I. R. (2005). Soil bulk density and crop yield under eleven consecutive years of corn with different tillage and residue practices in a sandy loam soil in central Canada. *Soil Till. Res.*, *84*, 41-53.

De Boodt, M. & De Leenheer, L. (1967). Aggregate stability determination in the field. In: De Boodt, M., De Leenheer, L., Frese, H., Low, A. J. & Peerlkamp, P. K. (Eds.), West European methods for soil structure determination. *Rijkslandbouwhogeschool Gent, Ghent*, pp. 60-61.

De Leenheer, L. (1959). Werkwijzen van de analysen aan het Centrum voor Grondonderzoek (in Dutch). Rijkslandbouwhogeschool Gent, Ghent.

De Leenheer, L & De Boodt, M. (1959). Determination of aggregate stability by the change in Mean Weight Diameter. Proceedings of the international symposium on soil structure. Ghent, Belgium. Rijkslandbouwhogeschool, Ghent, Belgium, pp. 290-300.

Debout, H. (2004). Erosiebestrijding: Probleemstelling, preventie en bestrijding van watererosie in de Belgische akkerbouw (in Dutch). Thesis, Katholieke Hogeschool Kempen, Geel.

Deen, W. & Kataki, P. K. (2003). Carbon sequestration in a long-term conventional versus conservation tillage experiment. *Soil Till. Res.*, *74*, 143-150.

D'Haene, K. (2008). *The potential of reduced tillage agriculture in Flanders*. Dissertation, Ghent University, Ghent.

D'Haene, K., Vermang, J., Cornelis, W. M., Leroy, B. L. M., Schiettecatte, W., De Neve, S., Gabriels, D. & Hofman, G. (2008). Reduced tillage effects on physical properties of silt loam soils growing root crops. *Soil Till. Res.*, *99*, 279-290.

D'Haene, K., Sleutel, S., De Neve, S., Gabriels, D. & Hofman, G. (2009). The influence of reduced tillage on carbon dynamics in silt loam soils. *Nutr. Cycl. Agroeecosys*, DOI 10.1007/s10705-008-9240-9.

Dolan, M. S., Clapp, C. E., Allmaras, R. R., Baker, J. M. & Molina, J. A. E. (2006). Soil organic carbon and nitrogen in a Minnesota soil as related to tillage, *residue and nitrogen management. Soil Till, Res.*, *89*, 221-231.

Drury, C. F., Tan, C. S., Waleky, T. W., Oloya, T. O., Hamil, A. S. & Weavor, S. E. (1999). Red clover and tillage influence on soil temperature. Water content and corn emergence. *Agron. J*, *19*, 101-108.

Duiker, S. W. & Beegle, D. B. (2006). Soil fertility distributions in long-term no-till, chisel/disk and moldboard plow/disk systems. *Soil Till. Res.*, *88*, 30-41.

Ekeberg, E. & Riley, H. C. F. (1996). Effects of mouldboard ploughing and direct planting on yield and nutrient uptake of potatoes in Norway. *Soil Till. Res.*, *39*, 131-142.

Elrick, D. E., Reynolds, W. D. & Tan, K. A. (1989). Hydraulic conductivity measurements in the unsaturated zone using improved well analyses. *Ground Wat. Monit. Rev.*, *9*, 184-193.

El Titi, A. (2001). Non-inversion tillage in integrated farming concepts: prospects and constraints of cropping systems in Southwest of Germany. In: García-Torres, L., Benites, J. & Martínez-Vilela, A. (Eds.), Conservation agriculture, a worldwide challenge. Volume I: Keynote contributions, Environment, farmers' experiences, innovations, socio-economy and policy. *Springer*, *Córdoba*, pp. 201-209.

Esteve, J. F., Imeson, A., Jarman, R., Barberis, R., Rydell, B., Sánchez, V. C. & Vandekerckhove, L. (2004). Pressures and drivers causing soil erosion. In: Van-Camp, L., Bujarrabal, B., Gentile, A. R., Jones, R. J. A., Montanarella, L., Olazabal, C. & Selvaradjou, S. K. (Eds.), Reports of the technical working groups established under "Thematic strategy for soil protection". Office for official publications of the European Communities, Luxembourg, EUR 21319 EN/2, pp. 133-149.

Etana, A., Håkansson, I., Zagal, E. & Bučas, S. (1999). Effects of tillage depth on organic carbon content and physical properties in five Swedish soils. *Soil Till. Res.*, *52*, 129-139.

Friedel, J. K., Munch, J. C. & Fischer, W. R. (1996). Soil microbial properties and the assessment of available soil organic matter in a haplic luvisol after several years of different cultivation and crop rotation. *Soil Biol. Biochem*, *28*, 479-488.

Gillijns, K., Govers, G., Poesen, J., Mathijs, E. & Bielders, C. (2005). Stand van zaken bodemerosie in België (in Dutch). Koninklijk Instituut voor het Duurzaam Beheer van de Natuurlijke Rijkdommen (KINT), Brussels.

Gobin, A. & Govers, G. (2003). Pan-European soil erosion risk assessment project (Pesera). Third annual report to the European Commission, EC Contract No QLK5-CT-1999-01323. European Environment Agency (EEA), Copenhagen.

Goyens, S., Schiettecatte, W., Verbist, K., Serlet, L., Verelst, M., Mahieu, J., Taverniers, V., Cornelis, W. M. & Gabriels, D. (2005). Impact of alternative tillage techniques on erosion and runoff during early crop development. In: Reorganizing field and landscape structures in a context of building strategies for water and soil protection, Book of abstracts of COST 634. *Institute of Agrophysics*, 17-18/09/2005, Lublin, pp. 14-15.

Hamza, M. A. & Anderson, W. K. (2005). Soil compaction in cropping systems review of the nature, causes and possible solutions. *Soil Till. Res.*, *82*, 121-148.

Hermawan, B. & Bomke, A. A. (1997). Effects of winter cover crops and successive spring tillage on soil aggregation. *Soil Till. Res.*, *44*, 109-120.

Hussain, I., Olson, K. R. & Ebelhar, S. A. (1999). Long-term tillage effects on soil chemical properties and organic matter fractions. *Soil Sci. Soc. Am.* J, *63*, 1335-1341.

Ide, G. & Hofman, G. (1990). The influence of subsoiling a plough-sole on the yield of agricultural crops. *Soil Technol.*, *3*, 259-268.

Ide, G., Hofman, G., Ossemerct, C. & Van Ruymbeke, M. (1984). Root growth of winter barley in response to subsoiling. *Soil Till. Res.*, *4*, 419- 431.

Jones, R. J. A., Le Bissonnais, Y., Bazzoffi, P., Díaz, S. J., Düwel, O., Loj, G., Øygarden, L., Prasuhn, V., Rydell, B., Strauss, P., Üveges, J. B., Vandekerckhove, L. & Yordanov, Y. (2004). Nature and extent of soil erosion in Europe. In: Van-Camp, L., Bujarrabal, B., Gentile, A. R., Jones, R. J. A., Montanarella, L., Olazabal, C. & Selvaradjou, S. K. (Eds.), Reports of the technical working groups established under "Thematic strategy for soil protection". Office for official publications of the European Communities, Luxembourg, EUR 21319 EN/2, pp. 150-190.

Kay, B. D. & Vanden Bygaart, A. J. (2002). Conservation tillage and depth stratification of porosity and soil organic matter. *Soil Till. Res.*, *66*, 107-118.

KMI. (2007). Statistieken (in Dutch and French) (accessed on 30/03/2007) http://www.meteo.be/nederlands/pages/Klimatologisch/century/statistieken.html

Larney, F. J., Ren, J., McGinn, S. M., Lindwall, C. W. & Izaurralde, R. C. (2003). The influence of rotation, tillage and row spacing on near-surface soil temperature for winter wheat in southern Alberta. *Can. J. Soil Sci.*, *83*, 89-98.

Layton, J. B., Skidmore, E. L. & Thompson, C. A. (1993). Winter-associated changes in dry-soil aggregation as influenced by management. *Soil Sci. Soc. Am. J*, *57*, 1568-1572.

Le Bisonnais, Y. (1996). Agreggate stability and assessment of soil crustability and erodibility: I. Theory and methodology. *Eur. J. Soil Sci.*, *47*, 425-437.

Le Bissonnais, Y., Cros-Cayot, S. & Gascuel-Odoux, C. (2002). Topographic dependence of aggregate stability, overland flow and sediment transport. *Agronomie*, *22*, 489-501.

Legout, C., Legue, S. & Le Bissonnais, Y. (2005). Aggregate breakdown dynamics under rainfall compared with aggregate stability measurements Eur. *J. Soil Sci.*, *56*, 225-237.

Leys, A., Govers, G., Gillijns, K. & Poesen, J. (2007). Conservation tillage on loamy soils: explaining the variability in interrill runoff and erosion reduction. *Eur. J. Soil Sci.*, *58*, 1425-1436.

Liebig, M. A., Tanaka, D. L. & Wienhold, B. J. (2004). Tillage and cropping effects on soil quality indicators in the northern Great Plains. *Soil Till. Res.*, *78*, 131-141.

Lipiec, J., Kuś, J., Slowińska-Jurkiewics, A. & Nosalewics, A. (2006). Soil porosity and water infiltration as influenced by tillage methods. *Soil Till. Res.*, *89*, 210-220.

Lupwayi, N. Z., Arshad, M. A., Rice, W. A. & Clayton, G. W. (2001). Bacterial diversity in water-stable aggregates of soils under conventional and zero tillage management. *Appl. Soil Ecol.*, *16*, 251-261.

Martens, D. A. (2000). Plant residue biochemistry regulates soil carbon cycling and carbon sequestration. *Soil Biol. Biochem*, *32*, 361-369.

Mathieu, C. & Pieltain, F. (1998). Analyse physique des sols: Méthodes choisies (in French). Lavoisier Tec. Doc. *París, London*, New York.

Matula, S. (2003). The influence of tillage treatements on water infiltration into soil profile. *Plant soil Environ.*, *49*, 298-306.

Mehdi, B. B., Madramootoo, C. A. & Mehuys, G. R. (1999). Yield and nitrogen content of corn under different tillage practices. *Agron. J*, *91*, 631-636.

Pardo, A., Amato, M. & Chiaranda, F. Q. (2000). Relationships between soil structure, root distribution and water uptake of chickpea (*Cicer arietinum L.*). Plant growth and water distribution. *Eur. J. Agron*, *13*, 39-45.

Perfect, E. & Caron, J. (2002). Spectral analysis of tillage-induced differences in soil spatial variability. *Soil Sci. Soc. Am. J*, *66*, 1587-1595.

Phylip, J. R. (1985). Aproximate analysis of the borehole permeameter in unsaturated soils. Water Resour. *Res.*, *21*, 1025-1033.

Pimentel, D., Harvey, C., Resosudarmo, P., Sinclair, K., Kurz, D., McNair, M., Crist, S., Shpritz, L., Fitton, L., Saffouri, R. & Blair, R. (1995). Environmental and economic cost of soil erosion and conservation benefits. *Science*, *267*, 1117-1123.

Powlson, D. S. (2007). Long-term experiments as research facilitates for studying soil organic matter dynamics and soil functioning. In: Chabbi, A. (Ed.), Proceedings of the international symposium "Organic matter dynamics in agro-ecosystems". Institut Scientifique de Recherche Agronomique (INRA), 16-19/07/2007, Poitiers, pp. 72-73.

Puget, P. & Lal, R. (2005). Soil organic carbon and nitrogen in a Mollisol in central Ohio as affected by tillage and land use. *Soil Till. Res.*, *80*, 201-213.

Reynolds, W. D. & Elrick, D. E. (1990). Ponded infiltration from a single ring: I. Analysis of steady state flow. *Soil Sci. Soc. Am. J*, *54*, 1233-1241.

Reynolds, W. D & Elrick, D. E. (1991). Determination of hydraulic conductivity using a tension infiltrometer. *Soil Sci. Soc. Am. J*, *55*, 633-639.

Reynolds, W. D. & Elrick, D. E. (2002). Falling head soil core (tank) method. In: Dane, J. H. & Topp, G. C. (Eds.), *Methods of soil analysis*. Part 4. *Physical methods*, SSSA Book Series No. 5. SSSA, Madison, pp. 809-812.

Riley, H. C. F. (1998). Soil mineral -N and N- fertilizer requirements of spring cereals in two long - term tillage trials on loam soil in southeast Norway. *Soil Till. Res.*, *48*, 265-274.

Riley, H. C. F., Bleken, M. A., Abrahamsen, S., Bergjord, A. K. & Bakken, A. K. (2005). Effects of alternative tillage systems on soil quality and yield of spring cereals on silty clay loam and sandy loam soils in the cool, wet climate of central Norway. *Soil Till. Res.*, 80, 79-93.

Roisin, C. J. C. (2007). A multifractal approach for assessing the structural state of tilled soils. *Soil Sci. Soc. Am. J*, *71*, 1-12.

Rücknagel, J., Hofman, B. & Christen, O. (2004). Effect of soil-tillage on soil physical properties, total organic carbon content and winter barley yield in a long-term experiment in Germany. In: Fischer, T., Turner, N., Angus, J., McIntyre, L., Robertson, M., Borrell, A. & Lloyd, D. (Eds.), Proceedings of the 4th International Crop Science Congress (ICSS) "New directions for a diverse planet". *The Regional Institute Ltd.*, 26/09-01/10/2004, *Brisbane*. http://www.cropscience.org.au/icsc2004/ (accessed on 05/05/2005).

Singh, B. & Malhi, S. S. (2006). Response of soil physical properties to tillage and residue management on two soils in a cool temperate environment. *Soil Till. Res.*, *85*, 143-153.

Six, J., Bossuyt, H., Degryze, S. & Denef, K. (2004). A history of research on the link between (micro)aggregates, soil biota, and soil organic matter dynamics. *Soil Till. Res.*, *79*, 7-31.

Soil Moisture Equipment Corp. (1992). Operating instructions for the tension infiltrometer attachments to the Guelph permeameter, Preliminary instructions. Soil Moisture Equipment Corp. *Santa Barbara*, CA, USA.

Soil Survey Staff. (1992). Soil survey laboratory methods manual. Soil Conservation Service. U.S. Dept. of agriculture. Rpt. N° 42. Washington. DC.

Stenberg, M., Stenberg, B. & Rydberg, T. (2000). Effects of reduced tillage and liming on microbial activity and soil properties in a weakly-structured soil. *Appl. Soil Ecol., 14*, 135-145.

Strauss, P., Swoboda, D. & Blum, W. E. H. (2003). How effective is mulching and minimum tillage to control runoff and soil loss? – a literature review. In: Gabriels, D. & Cornelis, W. M. (Eds.), Proceedings of the International Symposium '25 years of assessment of erosion', UGent, Gent, pp. 545-550.

Tebrügge, F. & Düring, R. A. (1999). Reducing tillage intensity- a review of results from a long term study in Germany. *Soil Till. Res., 53*, 15-28.

Vandergeten, J. P. & Roisin, C. (2004). Ploegloze teelttechnieken in de suikerbietenteelt (in Dutch and French). De technische gidsen van het Koninklijk Belgisch Instituut tot Verbetering van de Biet (KBIVB). KBIVB, Tienen.

Van der Lecq, L., Klopper, A. & Jackman, I. (2001). Carrying capacity of soil. Eijkelkamp. Agrisearch Equipment. Geijkt Nieuws.

Vermang, J., Serlet, L., Goyens, S., Schiettecatte, W., Cornelis, W., Mahieu, J. & Gabriels, D. (2007). Impact of conservation tillage on soil erosion and crop yields for maize in monoculture and for a maize – wheat – sugar beet rotation. In: proceedings of the 5th International Congress of the European Society for Soil Conservation (ESSC) "Changing soils in a changing world: the soils of tomorrow", 25-30/06/2007, Palermo, pp. 392.

Verstraeten, G. & Poesen, J. (1999). The nature of small-scale flooding, muddy floods and retention pond sedimentation in central Belgium. *Geomorphology, 29*, 275-292.

Verstraeten, G., Poesen, J., Goossens, D., Gillijns, K., Bielders, C., Gabriels, D., Rysschaert, G., Van Den Eeckhaut, M., Vanwalleghem, T. & Govers, G. (2006). 1.30. Belgium. In: Boardman, J. & Poesen, J. (Eds.), Soil Erosion in Europe. John Whiley & Sons Ltd., *Chichester*, pp. 385-411.

Verstraeten, G., Poesen, J., Govers, G., Gillijns, K., Van Rompaey, A. & Van Oost, K. (2003a). Integrating science, policy and farmers to reduce soil loss and sediment delivery in Flanders, Belgium. *Environ. Sci. Policy, 6*, 95-103.

Verstraeten, G., Van Rompaey, A., Van Oost, K., Govers, G. & Poesen, J, (2003b). Milieu- en natuurrapport Vlaanderen, Achtergronddocument 2.21b Kwaliteit Bodem: erosie (in Dutch). Vlaamse Milieumaatschappij (VMM), Erembodegem, p. 345-355.

Walkley, A. & Black, I. A. (1934). An examination of the Degtjareff method for determining soil organic matter, and a proposed modification of the chromic acid titration method. *Soil Sci., 34*, 29-38.

Wienhold, B. J. & Tanaka, D. L. (2000). Haying, tillage, and nitrogen fertilization influences on infiltration rates at a conservation reserve program site. *Soil Sci. Soc. Am. J, 64*, 379-381.

Yang, X. M. & Wander, W. M. (1999). Tillage effects on soil organic carbon distribution and storage in a silt loam soil in Illinois. *Soil Till. Res., 52*, 1-9.

In: No-Till Farming
Editor: Earl T. Nardali

ISBN: 978-1-60741-402-5

Chapter 2

EFFECTS AND AFTER-EFFECTS OF 20 YEARS OF REDUCED TILLAGE PRACTICES ON SOIL PROPERTIES AND CROP PERFORMANCE IN THE SEMI-ARID SUBTROPICS IN AUSTRALIA

***B.J. Radford*[A] *and C.M. Thornton*[A]**

[A]Queensland Department of Environment and Resource Management, LMB 1, Biloela, 4715 Queensland, Australia.

ABSTRACT

A long-term tillage experiment was designed to assess the effect of tillage frequency and intensity on rainfed grain production and quality in the semi-arid subtropical environment of central Queensland, Australia. There were four tillage treatments: traditional tillage (TT), stubble mulch tillage (SM), reduced tillage (RT) and no till (NT), each with and without applied fertiliser (N+Zn). On completion, after 20 years of treatment application, all treatments were managed using no till and appropriate fertiliser (N+Zn) application for a further 5 years.

During the 20 years when tillage treatments were being applied, the reduced tillage treatments (NT, RT and SM) outyielded TT in fewer than half of the 22 crops grown. Mean yields without fertiliser were 2.0 t/ha (TT) and 2.6 t/ha (NT) while mean yields with fertiliser were 1.9 t/ha (TT) and 2.9 t/ha (NT). During the subsequent 5 years of no till, the former reduced tillage treatments outyielded the former TT in each of the 5 crops grown. For example, the long-term NT gave an average yield of 3.3 t/ha while the short-term NT, which was formerly TT, produced only 2.1 t/ha - a 57% yield increase. This increase was due to both increased soil water storage and higher water use efficiency (WUE). Both were attributed to the development of improved soil structure, higher population densities of soil macrofauna and slightly higher soil organic carbon content. High WUE in NT was also attributed to a beneficial effect resulting from slow early growth under no till. Results indicate it takes at least 20 years to attain the full soil benefits (physical, chemical and biological) of a no-till system. The large yield responses from the three reduced tillage treatments, during and after treatment application, were realised because cropping frequency was higher than would have been appropriate for traditional tillage. Increased cropping frequency also results in higher levels of groundcover, which reduces soil erosion and creates a more sustainable farming system. A high-yielding, viable cropping system can also contribute towards environmental sustainability by reducing the need for further land clearing.

Keywords: no till, grain yield, soil water storage, organic carbon, compaction

INTRODUCTION

Reduced tillage practices in Australia were originally adopted to reduce soil erosion [Thomas et al., 2007] and therefore maintain long-term productivity. Initially many growers had difficulty weaning themselves off their traditional tillage practices because reduced tillage cost more and tended to yield less. Problems arose with fallow weed control, crop establishment, crop nutrition, and an increased incidence of pests and diseases. Even when technology was developed to overcome these problems, positive yield responses occurred only occasionally. Uren [1991] stated that "the remarkable thing about the introduction of minimum tillage systems is that throughout the world, irrespective of climate, soil type, crop and fertilizer practice, the result has been the same - very little change in yield." He added that "this result must be extremely disappointing to the proponents of minimum tillage systems." For example, an 11-year study in south east Queensland had shown that no till increased fallowing efficiency but not necessarily the yield of winter cereal crops [Marley and Littler, 1989]. Advocates of minimum tillage attributed the dearth of positive on-farm yield responses to weather conditions and failure to apply proven technology. A belief then began to develop that any soil response to the new management regime was going to be slow to develop.

The initial aim of the research reported here was to assess the effect of tillage frequency and intensity on rainfed grain production and quality in the semi-arid subtropical environment of central Queensland. This aim was achieved during 20 years of treatment application. The trial was then terminated and during the next 5 years the entire trial area was managed uniformly with no tillage + fertiliser application. During those 5 years, marked differences in crop performance were observed in response to the previous 20 years of tillage treatments, and an opportunity was identified to measure the residual effects. Grain yields and soil water contents at sowing and harvest were determined each year in the former plots to test the hypothesis that prolonged no tillage improves the condition of the soil. Results revealed consistent responses to the former treatments. The carry-over or residual effects exceeded the responses generally obtained from the conservation tillage treatments during the 20 years of treatment application. It is now evident that reduced tillage in time confers exciting yield responses in the semi-arid subtropics.

MATERIALS AND METHODS

Site

The site at Biloela (24.37°S, 150.52°E) was cleared in1924 and used for pasture and rainfed crop production until April 1983 when the tillage treatments commenced. The land is non-sloping.

The soil was developed on an alluvium and comprises black cracking and non-cracking clays, classified as Tognolini and Melton [Shields, 1989], respectively, or Vertisol and Aridisol [Soil Survey Staff, 1975], respectively. Details of some soil physical and chemical properties are given in Radford et al. [1995].

The 84-year mean annual rainfall is 675 mm (range 317-1132 mm), with 72% falling in summer (October-March). The mean annual evaporation from a class A pan is 1870 mm.

Treatments and Design

The four main treatments were first imposed in April 1983 immediately after growing a sorghum crop and were:

(1) Traditional tillage (TT) using a disc plough, scarifier and cultivator for all weed control
(2) Stubble mulch tillage (SM) using stubble mulching implements (chisel plough, blade plough and rod weeder) for all weed control
(3) Reduced tillage (RT) using stubble mulching implements or herbicides to control weeds
(4) No till (NT) using herbicides for all weed control.

TT had the most soil disturbance and NT (sowing tines only) the least (TT > SM > RT > NT). All crop residues were retained *in situ* but levels of ground cover varied with the treatments, due to different rates of incorporation by the tillage implements [Sallaway et al., 1988] and different amounts of residues produced. TT had the least ground cover (after inversion of the surface soil during the primary disc tillage operation) and NT the most (TT < SM < RT < NT). During the 20 years of treatment application, TT and SM each received 86 tillage operations during the fallow periods, RT 53 and NT none.

The experimental design was a randomised block with 4 tillage treatments and 4 replications. Fertiliser application commenced in 1989 when each main plot was split for 2 fertiliser treatments (control, N+Zn). N was applied as urea at rates (kg N/ha) of 100 (1989), 80 (1990), 120 (1991), 75 (1994), 50 (1997 and 1998), 40 (1999), 50 (2000) and 40 (2002). Zn was applied as 1-2 foliar sprays of 1% zinc sulphate heptahydrate when required, and as zinc sulfate monohydrate at 40 kg/ha in November 1996. Split-plot analysis of variance was carried out for all measurements. The main plots were 72 x 22 m.

All crops were sown at recommended seeding rates using a no-till planter equipped with smooth coulters preceding spearpoint openers and press wheels. Opportunity cropping was practised, a summer or winter crop being sown whenever any treatment had enough stored soil water to ensure satisfactory crop establishment and a high probability of a harvestable yield. Details of the crops grown and rainfall received are given in Table 1. When treatment application ceased (after the 2003 sorghum crop), 5 sorghum crops (including one ratoon crop) were grown with the same no-till management across the entire trial area. Measurements were continued in the split plots but fertiliser was applied as a blanket application to all plots: 40 kg N/ha (2005), 50 kg N/ha (2006), 70 kg N/ha (2007) and foliar sprays of zinc sulphate heptahydrate when required.

Table 1. Details of crops grown and rainfall totals during 20 years of treatment application followed by 5 years of no till with N+Zn fertiliser

Crop species	Cultivar	Sowing date	Harvest date	Fallow		Crop	
				Period (days)	Rainfall (mm)	Period (days)	Rainfall (mm)
Sorghum	E57	3.2.84	12.6.84	279	528	130	106
Sorghum	E57	17.12.84	26.4.85	188	448	130	313
Sorghum	E57	18.12.85	10.4.86	236	458	113	172
Sorghum	Prize	8.10.86	9.2.87	181	278	124	413
Wheat	Hartog	8.5.87	28.10.87	88	175	173	234
Wheat	Hartog	25.5.88	12.10.88	210	408	140	161
Wheat	Hartog	2.6.89	19.10.89	233	601	139	125
Wheat	Flinders	9.5.90	24.10.90	204	376	168	153
Wheat	Hartog	31.5.91	17.10.91	219	358	139	37
Wheat	Vasco	2.6.92	26.10.92	229	641	146	101
Sorghum	DK48M	25.1.93	26.5.93	91	241	121	62
Maize	C37	14.1.94	1.6.94	233	391	138	265
Maize	Hycorn 72	12.1.95	18.5.95	225	248	126	144
Wheat	Batavia	9.5.96	18.10.96	357	566	162	234
Mungbean	Emerald	26.11.96	28.1.97	39	72	63	93
Wheat	Sunvale	29.5.97	16.10.97	121	377	140	62
Wheat	Batavia 2	13.5.98	30.10.98	209	400	170	261
Wheat	Kennedy	18.6.99	15.10.99	231	618	119	46
Maize	DK689	19.1.00	27.6.00	96	333	160	171
Sorghum	MR31	4.1.01	29.5.01	191	288	145	183
Sorghum	MR43	24.12.01	18.4.02	209	294	115	165
Sorghum	MR43	17.2.03	1.7.03	305	683	134	215
Ratoon sorghum	MR43	29.8.03	20.1.04	59	128	144	374
Sorghum	MR43	11.1.05	25.5.05	357	430	134	211
Sorghum	MR43	17.11.05	13.3.06	176	294	116	317
Sorghum	MR43	20.11.06	16.3.07	252	364	116	197
Sorghum	MR43	17.10.07	18.2.08	215	225	124	180

Measurements

Soil bulk density. Bulk densities within the soil profile were measured in 1991 in order to calculate volumes of water in the soil. Known volumes of soil at field capacity were oven-dried and weighed. Eight depth increments (0-0.1, 0.1-0.2, 0.2-0.3, 0.3-0.5, 0.5-0.7, 0.7-1.0, 1.0-1.3 and 1.3-1.6 m) were sampled in each treatment. Infiltration rings were used to attain field capacity in the top three depth increments.

Soil water content. From 1989 to 1999, gravimetric soil water contents were determined at sowing and harvest of each crop using a hydraulically powered sampler to extract soil cores to a depth of 1.6 m. Two cores from randomly selected locations in each control and each fertilised subplot were taken, divided into the 8 depth increments listed above and dried to constant weight at 105°C.

From 2000 to 2008, soil water contents were determined at sowing and harvest of each crop by neutron moisture meter (Campbell Pacific Nuclear Model 503 DR Hydroprobe) to a depth of 1.5 m. Aluminium access tubes were installed in each subplot after sowing, and soil water contents were measured in 15 x 0.1 m increments. Total water in each profile was determined by addition.

Total soil water contents are presented because lower limits for each crop were not available.

Soil water accumulation during each fallow was determined from the difference between soil water at sowing and soil water at the previous harvest. Fallowing efficiency was then calculated as the percentage of fallow rainfall accumulated in the soil.

Soil nitrate nitrogen (NO_3-N) content. From 1991 to 2003, soil samples to determine the quantity of NO_3-N in the root zone were taken at sowing on 10 occasions. Two cores from each subplot were cut into the 8 depth increments with measured bulk densities, bulked, dried at 40°C, ground to pass through a 2 mm sieve, and analysed for NO_3-N by KCL extraction. Quantities at 0-1.6 m were calculated using the bulk density data.

Soil penetration resistance. Cone index was measured in all plots in May 1997 (after 14 years of treatment application using) using a hydraulically powered recording cone penetrometer mounted on a tractor. Values were obtained in 15 mm increments to a depth of 600 mm when the soil was at the upper drainage limit. A standard 30° circular stainless steel cone of 12.83 mm diameter [ASAE Standard S313.3, 2004] was used, and three probes were taken in each subplot.

Soil organic carbon content. Five soil cores to a depth of 0.2 m were taken at random locations within each subplot on 13 Apr 1989 (prior to any fertiliser application), 14 May 1997 and 15 Sep 2004. Depth increments of 0-0.1 and 0.1-0.2 m were separated, bulked, dried at 40°C, finely ground to pass through a 0.5 mm sieve, thoroughly mixed, and analysed using the dichromate oxidation method [Walkley and Black, 1934] with titrimetric finish. The 1989 and 1997 samples were analysed in the same laboratory at the same time to accurately determine any change in organic carbon during that 8-year period. The 2004 samples provided an indication of the organic carbon status of the soil after 20 years of tillage treatments. Five cores at 0-0.1 m were also taken in the fertilised TT and NT subplots on 3 Apr 2008 and analysed as described above.

Crop establishment. The planter was calibrated to determine the number of seeds sown per unit length of row, and the number of emerged seedlings was counted in each crop 7-28 days after sowing in a total row length of 2.4-50 m per subplot. Establishment percentages were calculated as a percentage of the number of seeds sown, and plant populations were also determined.

Grain yield. Grain was harvested from measured areas of 25-200 m^2 in each subplot using a small-plot harvester, and weighed. Grain moisture contents were measured with a moisture meter, and all yields were adjusted to12% moisture content.

Aboveground dry matter after harvesting. The mean annual mass of aboveground dry matter returned to the soil was calculated using harvest indices of 52% for sorghum and maize, 50% for mungbean and 39% for wheat [Bradford et al., 2005].

Grain yield components. Mean grain weight was determined from1-2 lots of 100 seeds per subplot, and the mean number of grains per m^2 was calculated.

Grain N content. Grain N in each subplot was determined for 17 crops between 1988 and 2004. Grain samples were analysed for N content using the Kjeldahl method [Bradstreet, 1965] with a flow injection automated colorimetric finish. Wheat protein contents were determined from grain N content at 0% moisture x 5.7 [Tkachuk, 1969].

Crop water use efficiency (WUE). WUE for grain production was determined for each crop using grain yield / (soil water accumulation during the fallow + in-crop rainfall) [Perry, 1987].

RESULTS

Soil Water Storage and Use

During the period of tillage treatment application, 7 of 16 crops (44%) had significant treatment differences in soil water content at sowing (data not shown). On each occasion, one or more of the three conservation tillage treatments (SM, RT and NT) contained more stored water than TT. After treatments ceased, all 5 subsequent crops (100%) had significant differences among the four former treatments. During the treatment period, SM, RT and NT had an extra 41-50 mm compared with TT, and afterwards they had an extra 65-67 mm (Table 2). Soil water use was also higher in the conservation tillage treatments, compared with TT, both during and after treatment application (Table 2). Even 5 years of NT following 20 years of TT were insufficient to attain the amount and depth of soil water use of 25 years of NT (Figure 1). Fallowing efficiencies were generally low (Table 2), due in part to the fact that soil water accumulates before harvesting and is most efficient at this time because the soil profile is dry.

Soil Nitrate Nitrogen (NO_3-N) Accumulation

During the 20 years when tillage treatments were being applied, TT had significantly ($P<0.05$) more NO_3-N than SM and RT, both of which had significantly more than NT (Table 2). This response occurred in both the unfertilised treatments and those which received applied N fertiliser (Table 2).

Table 2. Effect of four tillage treatments on soil properties and crop performance during 20 years of treatment application followed by 5 years of no till with N+Zn

Measurement	Period/time when measured	Fertiliser treatment	Treatment				lsd (P=0.05)
			TT	SM	RT	NT	
Total soil water at sowing (mm)	During trial (16 crops)	Various	485	526	535	531	11
	After trial (5 crops)	N+Zn	449	514	514	516	20
Soil water use (mm)	During trial (16 crops)	Various	95	119	124	122	12
	After trial (5 crops)	N+Zn	97	133	122	121	27
NO3-N at sowing (kg/ha)	During trial (10 crops)	Nil	66	57	52	40	7
		N+Zn	154	102	93	74	20
Fallowing efficiency (%)	During trial (15 crops)	Various	21	29	31	29	6
	After trial (5 crops)	N+Zn	42	55	46	43	9
Cone index: 0.36 m (MPa)	During trial (May 1997)	Various	2.00	1.81	1.68	1.51	0.29
Crop establishment (%)	During trial (22 crops)	Various	64	64	64	66	ns
Soil organic carbon (%):							
0-0.1 m	After trial (Sep 2004)	Various	1.37	1.60	1.54	1.61	ns
0.1-0.2 m	After trial (Sep 2004)	Various	1.29	1.45	1.35	1.41	0.10
Organic C decline (kg/ha):							
0-0.2 m	Apr 1989 - May 1997	Various	3111	1046	1378	255	626
Aboveground dry matter returned to the soil (t/ha)	20-year trial period	Nil	2.19	2.63	2.81	2.75	0.54
	(22 crops)	N+Zn	2.43	3.02	3.07	3.11	0.61
Grain yield (t/ha)	During trial (22 crops)	Nil	2.03	2.46	2.63	2.60	0.22
		N+Zn	1.94	2.72	2.71	2.85	0.19
	After trial (5 crops)	N+Zn	2.08	3.24	3.32	3.30	0.46
Grain N content (%)	During trial (14 crops)	Nil	1.94	1.67	1.57	1.43	0.08
		N+Zn	2.33	2.12	1.99	1.84	0.11
Grain N yield (kg/ha)	During trial (14 crops)	Nil	79	91	93	87	13
		N+Zn	115	150	136	131	13
Water use efficiency (kg/ha.mm)	During trial (16 crops)	Nil	7.1	8.7	9.4	9.4	1.0
		N+Zn	8.3	9.9	10.1	10.3	0.9
	After trial (5 crops)	N+Zn	6.3	8.6	9.1	9.0	1.0

Soil Strength

After 14 years of treatment application, NT had significantly ($P<0.05$) higher cone indices than TT at shallow depths (0.075-0.105 m) but significantly lower values deeper (0.33-0.39 m) (Figure 2). Similar but non-significant trends were apparent to a depth of 0.6 m, below which no further readings were taken. Since cone index increased with depth, the higher values in TT at depth (0.33-0.39 m) were most likely to affect soil water storage and crop root growth.

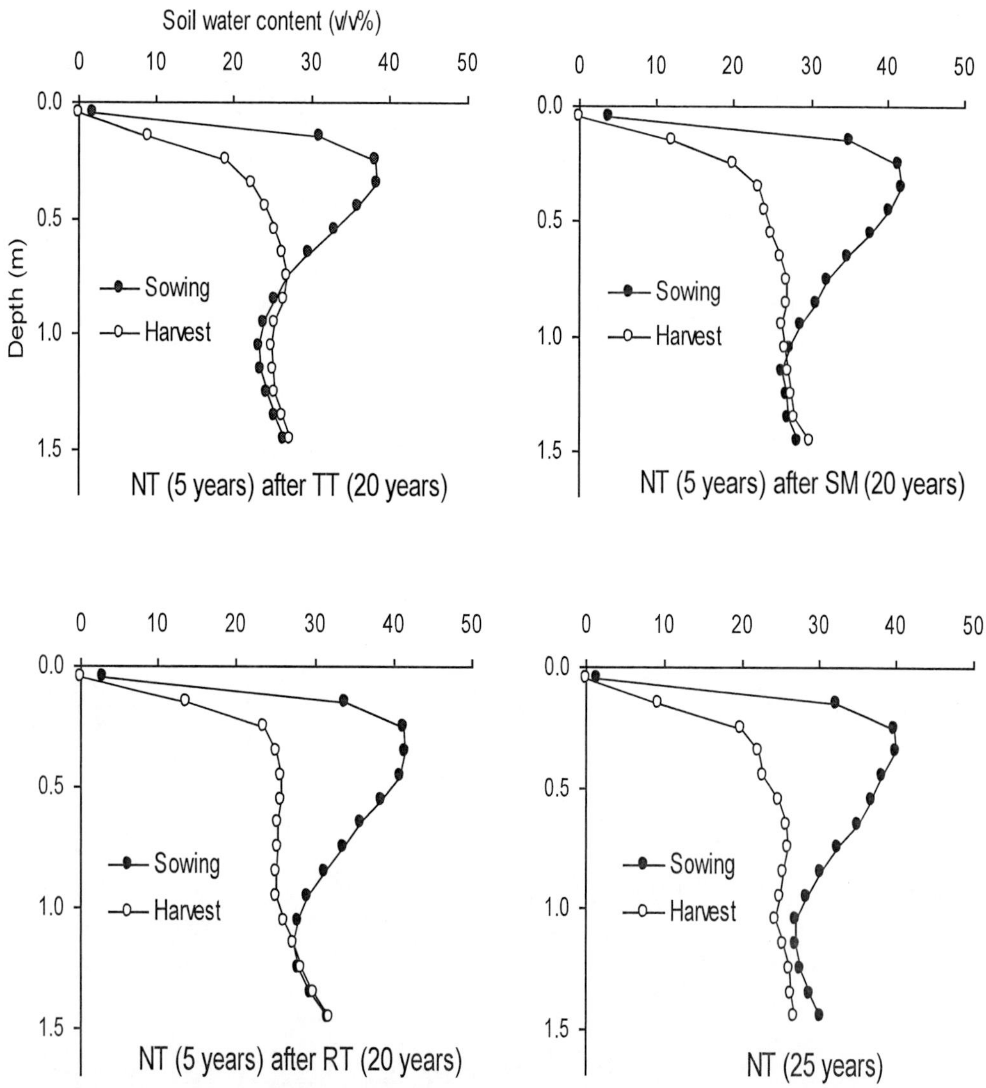

Figure 1. Soil water use between sowing and harvest after 5 years of no-till and 20 years of application of four tillage treatments (NT = no tillage; RT = reduced tillage; SM = stubble mulch tillage; TT = traditional tillage)

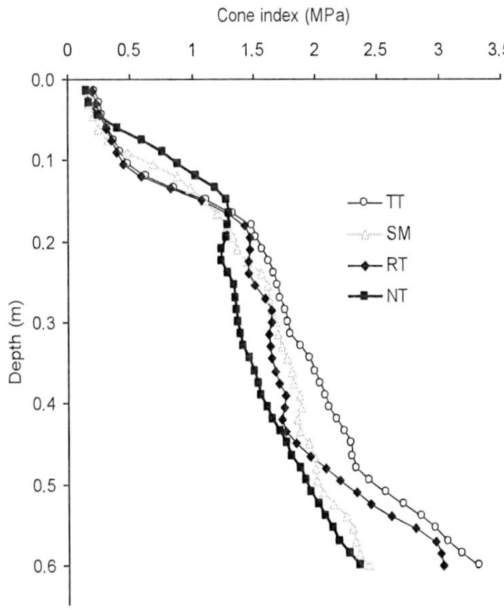

Figure 2. Effect of four tillage treatments on cone index from 0 to 0.6 m in May 1997 (NT = no tillage; RT = reduced tillage; SM = stubble mulch tillage; TT = traditional tillage)

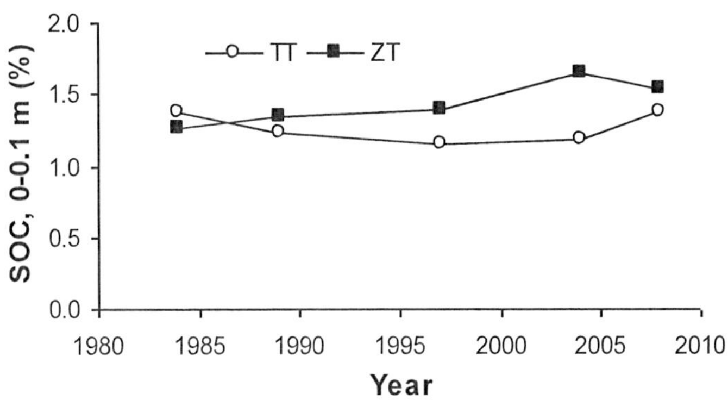

Figure 3. Soil organic carbon content at 0-0.1 m in fertilised NT and TT in 1984, 1989, 1997, 2004 and 2008 (NT = no tillage; TT = traditional tillage)

Soil Organic Carbon Content

Twenty years of no till resulted in no change in soil organic carbon at 0-0.1 m compared with traditional tillage and only a small increase at 0.1-0.2 m (Table 2). During the eight years from Apr 1989 to May 1997 the amount of organic carbon at 0-0.2 m declined in all treatments, but the decline was significantly greater in SM and RT compared with TT, and in NT compared with SM and RT (Table 2). The organic carbon content of the fertilised TT and NT plots at 0-0.1 m from 1984 to 2008 is shown in Figure 3. It tended to decline in TT but increase in NT during the period of the trial (1984-2004).

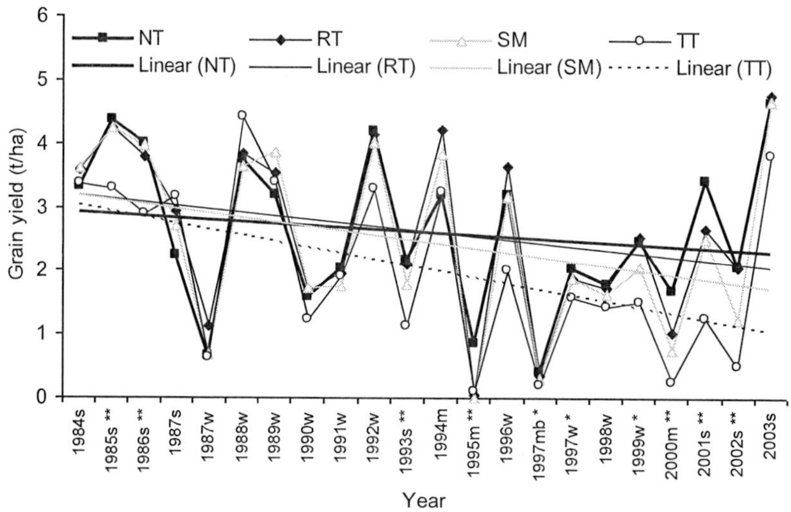

Figure 4. Grain yields without fertiliser during the 20-year trial and linear trends (NT = no tillage; RT = reduced tillage; SM = stubble mulch tillage; TT = traditional tillage; s = sorghum; w = wheat; m = maize; mb = mung bean; * = $P < 0.05$; ** = $P < 0.01$)

Crop Establishment

Mean crop establishment during the 20-year trial did not vary among the four tillage treatments (Table 2) but there were differences in individual sowings. NT had lower sorghum establishment than the other three treatments in 1984 (the first year of the trial) and lower maize establishment than SM and TT in 1995. On the other hand, NT had higher wheat establishment than TT in 1991 and 1992 and higher sorghum establishment than TT in 2002. Despite these occasional differences, plant population densities always lay within, or close to, the recommended range and any variations were unlikely to have affected grain yield.

Grain Yield

Mean yield during the 20-year trial was higher in NT, RT and SM than TT by 0.4-0.6 t/ha without fertiliser and by 0.8-0.9 t/ha with N+Zn (Table 2). Significant ($P<0.05$) differences occurred in 10 of 22 crops (45%) without fertiliser and in 4 of 16 crops (25%) with N+Zn. Similar, but non-significant, trends were apparent in the other crops. Yields varied widely but the linear trend lines in all the unfertilised tillage treatments declined. This decline was small for NT (due to drier conditions) but increased successively for RT, SM and TT (Figure 4). When all treatments became no till, long-term NT outyielded short-term NT (following 20 years of TT) each year for five years (Figure 5). The non-convergence of the trend lines in Figure 4 indicates that the yields of the short and long-term NT were showing no tendency to equalise after the five years. There was also no tendency for the soil water storage or WUE values in short and long-term NT to become closer during the five years (data not shown).

Mean yield during the 5 subsequent years of no till with fertiliser was lower in the former TT compared with NT, RT and SM by 1.2 t/ha (Table 2). Significant (P<0.05) yield differences occurred in all five crops (i.e. in 100% of post-trial crops for 5 years).

Aboveground Dry Matter after Harvesting

The mean annual input of aboveground dry matter in the unfertilised soil during the 20-year trial was higher in RT (2.81 t/ha) and NT (2.75) than in TT (2.19) (Table 2). The treatments that were fertilised after year 6 also received more aboveground dry matter per annum in RT (3.07) and NT (3.11) than in TT (2.43) (Table 2).

Grain N Content and Grain N Yield

Grain N content decreased in the order TT > SM > RT > NT, both with and without N fertiliser during the trial (Table 2). The total amount of N exported in the grain was higher in NT, RT and SM than TT, both with and without N fertiliser (Table 2).

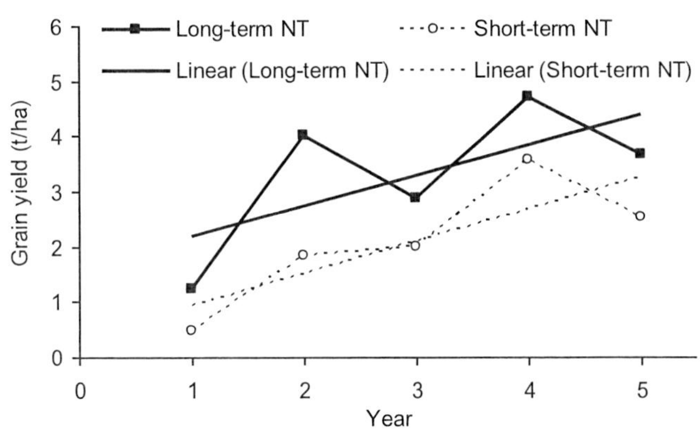

Figure 5. Effect of long and short-term NT on sorghum grain yield during five successive years. The long-term NT had 20 years of prior NT; the short-term NT had 20 years of prior TT. The former RT and SM treatments performed similarly to the long-term NT and are not shown (NT = no tillage; RT = reduced tillage; SM = stubble mulch tillage; TT = traditional tillage)

Crop WUE

WUE was higher in NT, RT and SM compared with TT both during and after the trial (Table 2). This means that NT, RT and SM not only stored more soil water than TT but used all their stored water more efficiently.

Discussion

Soil Water Storage

We found increased soil water storage in response to reductions in the intensity and frequency of tillage during the preceding fallow. Increased soil water storage following reduced tillage has also been reported elsewhere in central Queensland [Rawson *et al.*, 1981; Thomas *et al.*, 1990], in Australia [Marley and Littler, 1989; Felton et al., 1995] and in the USA [Barnes et al., 1955; Smika and Wicks, 1968; Unger, 1977]. Furthermore, there were more consistent responses after 20 years even when all treatments were managed uniformly using no till. Increased water storage in the soil improves crop yields and makes cropping more reliable but the additional harvested water may also reduce overland flow. This means there may be less water available for farm dams.

Improved soil water storage under reduced tillage has been attributed mainly to increased ground cover through stubble retention [Marley and Littler, 1989; Thomas *et al.*, 1990] and improved soil structure [McGarry *et al.*, 2001]. Clarke [1987] stated that "tillage causes deterioration of soil physical properties due to the pulverising and smearing effects of implements, and pressure from load-bearing wheels, resulting in some form of compaction and a marked reduction in infiltration capacity". Faulkner [1943] showed great foresight in pointing out that the use of the "plow" has actually destroyed the productiveness of soils.

At our site, NT had higher rates of sorptivity and hydraulic conductivity than TT, while high-energy simulated rain resulted in a highly visible surface seal on the bare TT soil [McGarry *et al.*, 2001]. In addition to these direct effects, tillage buries crop residues to expose the surface soil to raindrop impact. In the absence of residues on the soil surface, even no till has poor water storage [Thomas *et al.*, 1990; Radford *et al.*, 1992].

Figure 6. Increased infiltration of 125 mm of rainfall in a no till plot compared with a traditional tillage plot after 8 years of treatment application

Water Use Efficiency

Although water is the main factor limiting crop production in most regions of the semi-arid sub tropics, increases in soil water have not always resulted in higher grain yields. Reduced WUE in response to reduced tillage has been attributed to reduced soil nitrate concentrations [Radford *et al.*, 1992; Felton *et al.*, 1995] and to surface stratification of immobile nutrient elements [Cowie *et al.*, 1996; Wright *et al.*, 2007]. Reduced WUE in wheat after no till has resulted from slow early growth [Chan et al., 1987; Kirkegaard *et al.*, 1994] and increased incidence of yellow spot (caused by *Pyrenophora tritici-repentis*) [Rees and Platz, 1979; Marley and Littler, 1989; Felton *et al.*, 1995], crown rot (caused by *Fusarium graminearum*) [Wildermuth *et al.*, 1997], rhizoctonia root rot (caused by *Rhizoctonia solani* Kühn) [Roget *et al.*, 1996] and root lesion nematode (*Pratylenchus thornei*) [Marley and Littler, 1989].

Any reductions in crop WUE at our site in response to reduced tillage generally occurred because of an inadequate supply of soil N in the unfertilised treatments. Adequacy of N nutrition for grain production is indicated by a wheat grain N content exceeding 2.0% or 11.4% protein [Russell, 1963; Goos *et al.*, 1982; Strong and Holford, 1997]. No wheat crop grown under NT without N fertiliser between 1988 and 1999 exceeded this critical value but 6 of 8 wheat crops which received N+Zn fertiliser did so, indicating that our N fertiliser applications were generally adequate.

When the soil N supply was adequate, we generally found increased WUE in the three reduced tillage treatments, and this persisted even after the 20 years of treatment application. The agronomic problems which cause low WUE under reduced tillage were rare or of little consequence in our trial. There was nil to low incidence of root and foliar diseases, and a soil test showed root lesion nematodes were entirely absent. Central Queensland has the advantage of generally minimal incidence of disease in its major crops (wheat and sorghum) including the diseases associated with reduced tillage practices. Although slow early growth under no till was often apparent, it tended to increase grain yields, as found by Cornish and Lymbery [1987]. Surface stratification of nutrients did not occur in our no-till treatment [Cowie *et al.*, 1996]. High population densities of earthworms under no-till (17 m^{-2}) [Wilson-Rummenie *et al.*, 1999] may have reversed this process by "plowing" the soil [Darwin, 1881]. Termite activity at this site [Holt *et al.*, 1993; Wilson-Rummenie *et al.*, 1999] may also have reversed the process of stratification. Even when no till does result in stratification, making immobile nutrient elements largely inaccessible to crop roots, the undisturbed network of vesicular arbuscular mycorrhizhae (VAM) in no till [Kabir, 2004] is able to enhance nutrient uptake from the deeper soil. As a result of all these processes, the high WUE inherent in reduced tillage practices can be expressed.

So why did reduced tillage increase WUE once most of the usual agronomic constraints associated with tillage reduction were absent? Firstly, slow early growth was observed in the no-till treatment but this meant less soil water was used for vegetative growth and more for grain production, compared with TT, resulting in more efficient water use. Cornish and Lymbery [1987] found that slow early growth under no till could increase grain yield when dependence on stored soil water was high. Secondly, the presence of stubble on the soil surface would have prevented the surface sealing that occurred in TT, inhibiting the infiltration of in-crop rainfall. Thirdly, reduced soil compaction at depth may have hastened root growth [Radford et al. 2001a]. Hazelton and Murphy [2007] indicate that cereal root growth is greatly restricted at penetrometer resistances of 1.25-2.00 MPa, and values recorded in this study lay within, or exceeded, this range below 0.2 m. Finally, the presence of old root channels, earthworm holes and termite galleries at this site would have provided easy access for crop roots in nutrient-rich zones resulting in efficient root growth and high WUE.

Grain Production

Our results show that the three reduced tillage practices (NT, RT and SM) increased both soil water storage and crop WUE compared with TT. So not only was additional water stored in the profile but every millimetre of the total supply of stored water + in-crop rainfall was used more efficiently to produce large relative differences in yield. During the 20-year trial, NT outyielded TT by 28% without fertiliser and by 47% with N+Zn. In the 5 years of NT with N+Zn after the trial, the long-term NT outyielded the short-term NT (formerly TT) by 59%. Further increases are possible with no till. For example, if agronomic or genetic research can improve WUE, it will be doing so for a larger store of soil water. Similarly, if more water can be stored, it will all be used more efficiently. Such improved performance with conservation tillage can help to "drought-proof" grain farms and generally enhance economic sustainability. Environmental sustainability is also enhanced because the need for further land clearing is reduced, and forests and wildlife can be saved [Center for Global Food Issues, 2008; Avery, 1998, 2007; Borlaug, 2007; Green et al., 2005].

High annual yields can be achieved in reduced tillage systems by increasing the frequency of cropping in order to take full advantage of the additional water being stored during each fallow period. Yield per unit time is the relevant measure rather than yield per crop on a fixed rotation. By sowing a crop (winter or summer) whenever there is enough water stored in the soil to ensure reliable establishment and early crop growth - the practice of opportunity cropping [Freebairn et al., 2006] - advantage can be taken of the inherently higher WUE when using in-crop rainfall rather than fallow rainfall. More frequent cropping means more groundcover and hence more stored water, effects which can become additive. Increased groundcover and water use also reduce runoff and soil erosion [Freebairn and Wockner, 1986; Berndt and White, 1976]. More frequent cropping in combination with no-till is also likely to result in higher levels of soil organic matter. Furthermore, higher cropping frequency with no till helps avoid the increased deep drainage associated with no till [Turpin *et al.*, 1999; McGarry *et al.*, 2000], which wastes water that could otherwise be used for crop growth.

The risk of crop failure traditionally associated with opportunity cropping [Hayman *et al.*, 1996] will be greatly reduced when long-term no till reproduces the yield responses we have reported here. There were three crop failures (grain yield <0.5 t/ha) in the unfertilised TT in the latter years of our trial (2000, 2002 and 2004) while in those same years the unfertilised NT produced harvestable yields. The three failed TT crops averaged 0.4 t/ha and the three successful NT crops 1.68 t/ha.

Figure 7. Sorghum crop in long-term (22 years) and short-term (2 years) no-till plots

Long v. Short-Term No Till

Why did the newly established no-till treatment (following TT) perform worse than the long-term no till for five years? Clearly, time was needed for complete repair of the soil after the damage inflicted by 20 years of tillage operations. On a similar soil type nearby, applied soil compaction persisted for the same period of time (5 years) under no till management [Radford *et al.*, 2007]. If compaction persisted for such a time span, macrofauna populations would have remained low [Radford *et al.*, 2001b], thereby minimising their contribution to soil structural improvement. Douglas (1987) found that soil macrofauna take about 7 years to become abundant and diverse in temperate Australia after cultivation has ceased. The higher levels of soil organic carbon in the long-term no till (Table 2) may also have helped increase yield by increasing the supply of soil nutrients.

Can the onset of the beneficial effects of long-term no till be hastened? If soil compaction is the issue, roots of some crop species may be better able to loosen the compacted soil than others. For example, Pillai and McGarry [1999] reported that lablab and mungbean produced smaller peds and increased interconnected pore space more rapidly than either sorghum or wheat. Henderson [1989] found that lupin can serve as a biological plough. If earthworms are lacking, suitable species could be introduced during the early stages of no-till management.

Research in northern China has shown that continuous long-term (15 years) conservation tillage improved soil structure and biological activity [Li *et al.*, 2007]. This included significantly increased aggregate stability in the larger size classes, and greater capillary porosity (pores of diameter <60 μm). Canadian research showed that 20 years of continuous no till with full stubble retention but no nitrogen application outyielded 2 years of no-till [Derpsch, 2008]; the 20-year no till produced 2.8 t/ha and 2-year no-till 1.8 t/ha. Increase in soil carbon content was considered the main factor leading to this result, yet we obtained similar responses but little difference in soil carbon contents. Benefits from no till accrue over time, and Sá [2004] has recognised 4 phases in the evolution of a long-term no-till system: an initial phase (0-5 years), a transition phase (5-10 years), a consolidation phase (10-20 years) and a maintenance phase (> 20 years). It is only in the maintenance phase that maximum soil benefits (physical, chemical and biological) are achieved. We too found that 20 years were required to maximise the benefits of a no-till system. Derpsch [2008] claims that any tillage performed during the earlier phases means a return to the initial phase. In other words, farmers who till their soil occasionally will never realise the full benefits of a no-till system. This claim conflicts with our finding that reduced and stubble mulch tillage generally performed as well as no till. Occasional strategic tillage or even control of weeds solely by stubble mulch tillage did little harm in the short to medium term. Strategic tillage may be necessary to manage weeds that are difficult to control with herbicides, weeds that are too advanced to control with herbicides and herbicide-resistant weeds. Strategic tillage may also be needed to overcome surface stratification and therefore isolation of nutrient elements, or to facilitate rainfall infiltration when stubble is lacking. Several aims can be achieved with a single strategic operation, which should be carried out at a time when the risk of soil erosion is low. It follows that such occasional tillage operations will require the retention on-farm of some tillage equipment.

In order to find reasons for the responses obtained, our trial was included in a study by Bell *et al.* [2006] investigating biotic constraints to crop growth. They found there were widespread biotic constraints to crop growth in northern Australia but they were unrelated to the presence of any known pathogen and did not appear to be consistently affected by management. Samples from our trial were taken after 20 years of treatment application (on 21 Feb 2003).

Soil Organic Carbon

The rate of decline of soil organic carbon on cropping land in central Queensland is rapid, particularly during the hot wet summers. No till slowed the rate of decline compared with the tilled treatments (Table 2; Figure 3), as also found by Standley *et al.* [1990] in central Queensland. According to Dalal and Probert [1997], soil organic carbon declines under continuous cropping for at least two reasons: rates of decomposition are increased due to cultivation, and annual inputs are lower under a cropping system than they were under a grassland or forest system. Annual aboveground inputs were higher in RT and NT than in TT (Table 2), and similar differences probably occurred below the ground. RT and NT returned 11-14 t/ha more dry matter to the soil than TT during the 20 years when treatments were being applied. The continuing improvement in soil properties we have observed under a long-term no-till cropping regime may eventually sustain, or even increase, the level of organic carbon in the soil.

Adoption of No-Till Farming

Fortunately, technological progress has now made the large-scale adoption of no-till farming systems easier and more effective. Controlled traffic farming with GPS guidance systems [Yule, 1998] has made herbicide application more precise, resulting in less wastage of chemicals and less damage to the environment. Adoption of no till in central Queensland has increased markedly in the last decade. A survey of 120 growers showed their cropped land under no till increased from 8% in 1997 to 30% in 2001, and it was estimated to increase to 47% in 2005 [Kuskie *et al.*, 2003]. ABS data (Anonymous, 2001; 2003) for central Queensland show the percentage of crop area under no till increased from 5% in 1996 to 33% in 2001 to 38% in 2003. Similar trends have occurred elsewhere in Queensland and Australia [Thomas *et al.*, 2007]. Our results indicate that the full benefit to recent adopters – providing they continue the practise of no till - will be realised only in the 2020s.

CONCLUSION

No till provides a silver bullet for grain yield improvement by means of soil improvement, at least in central Queensland. Complete recovery of soil physical, chemical and biological properties following a prolonged period of intensive cultivation may take at least 20 years. The long-term benefits of no till may become increasingly important for cropping land where rainfall is declining as a result of climate change. Reduced tillage practices (NT, RT and SM) increased both soil water storage and crop WUE compared with TT. This was attributed to improved soil structure in the surface and subsoil, increased soil macrofauna populations, and slightly increased soil organic carbon content. The resulting yield benefits are best realised by increasing cropping frequency, which increases groundcover and water use, reduces soil erosion and creates a more sustainable farming system. A high- yielding system enhances economic sustainability and also environmental sustainability by reducing the need for further land clearing.

REFERENCES

Anonymous (2001). *Australian Bureau of Statistics* (2001): Agricultural census – selected shire results. Canberra, Australia: Australian Bureau of Statistics, 2000/01.

Anonymous (2003). *Australian Bureau of Statistics* (2003): 'AgStats 2002.' CD-ROM. Canberra, Australia: Australian Bureau of Statistics.

ASAE Standard S313.3 (Feb 2004): *Soil cone penetrometer*. 2950 Niles Road, St. Joseph, Michigan 49085, U.S.A.: American Society of Agricultural and Biological Engineers.

Avery, D. T. (1998). The importance of high farm yields to wildlife conservation. *Australian Sugar Cane,* Aug-Sep 1998, 30-32.

Avery, D. T. (2007). How high-yield farming saves nature. *Society*, *44*, 137-143.

Barnes, O. K., Bohmont, D. W. & Rauzi, F. (1955). Effect of chemical and tillage summer fallow upon water-infiltration rates. *Agronomy Journal*, *47*, 235-236.

Bell, M., Seymour, N., Stirling, G. R., Stirling, A. M., Van Zwieten, L., Vancov, T., Sutton, G. & Moody, P. (2006). Impacts of management on soil biota in Vertosols supporting the broadacre grains industry in northern Australia. *Australian Journal of Soil Science, 44*, 433-451.

Berndt, R. D. & White, B. J. (1976). A simulation-based evaluation of three cropping systems on cracking-clay soils in a summer rainfall environment. *Agricultural Meteorology, 16*, 211-229.

Borlaug, N. E. (2007). Sixty-two years of fighting hunger: personal recollections. *Euphytica*, 157, 287-297.

Bradford, J. B., Lauenroth, W. K. & Burke, I. C. (2005). The impact of cropping on primary production in the U.S. Great Plains. Ecology, 86, 1863-1872. Appendix A: Values and references for allocation ratios and harvest indices for agricultural crops in the U.S. *Great Plains are available in ESA's Electronic Data Archive:* Ecological Archives E086-098-A1.

Bradstreet R. B. (1965). The Kjeldahl method for organic nitrogen (1965). New York: *Academic Press.*

Center for Global Food Issues (2008). Growing more per acre leaves more land for nature. Available from: www.highyieldconservation.org/contill.html

Chan, K. Y., Mead, J. A. & Roberts, W. P. (1987). Poor early growth of wheat under direct drilling. *Australian Journal of Agricultural Research, 38*, 791-800.

Clarke, A. L. (1987). The need for fertility restoration/maintenance. In: J.P. Thompson & J.A. Doughton (Eds.), Proceedings of the Queensland Crop Production Conference 1987 (7-17). Brisbane, Australia: *QDPI Conference and Workshop Series,* QC 87007.

Cornish, P. S. & Lymbery, J. R. (1987). Reduced early growth of direct drilled wheat in southern New South Wales: causes and consequences. *Australian Journal of Experimental Agriculture, 27*, 869-880.

Cowie, B. A., Hastie, M., Hunt, S. B., Asghar, M. & Lack, D. W. (1996). Surface soil nutrient distribution following zero tillage and traditional tillage management. In: M. Asghar (Ed.), *Proceedings of the 8th Australian Agronomy Conference* (160-163). Toowoomba, Australia: The Australian Society of Agronomy, Inc.

Dalal, R. C. & Probert, M. E. (1997). Soil nutrient depletion. In: A. L. Clarke & P. B. Wylie (Eds.), *Sustainable Crop Production in the Sub-tropics* (49). Brisbane, Australia: QDPI Information Series QI97035.

Darwin, C. R. (1881). *The formation of vegetable mould through the action of worms, with observations on their habits*. London: John Murray.

Derpsch, R. (2008). No-tillage and conservation agriculture: a progress report. In: T. Godard, M.A. Zoebisch, Y.T. Gan, W. Ellis, A. Watson & S. Sombatpanit (Eds.), *No-till Farming Systems Special Publication No. 3* (7-39). Bangkok: World Association of Soil and Water Conservation.

Douglas, L. A. (1987). Effects of cultivation and pesticide use on soil biology. In: P.S. Cornish & J.E. Pratley (Eds.), *Tillage – New Directions in Australian Agriculture* (308-317). Brisbane, Australia: Inkata.

Faulkner, E. H. (1943). *Plowman's Folly*. University of Oklahoma, U.S.A.

Felton, W. L., Marcellos, H. & Martin, R. J. (1995). A comparison of three fallow management strategies for the long-term productivity of wheat in northern New South Wales. *Australian Journal of Experimental Agriculture, 35*, 915-921.

Freebairn, D. M., Cornish, P. S. Anderson, W. K. Walker, S. R. Robinson, J. B. & Beswick, A.R. (2006). Management systems in climate regions of the world-Australia. In: G.A. Peterson, P. W. Unger & W. A. Payne (Eds.), *Dryland Agriculture* (2nd Edition). Madison, WI: American Society of Agronomy, Crop Science Society of America, Soil Science Society of America.

Freebairn, D. M. & Wockner, G. H. (1986). A study of soil erosion on vertisols of the eastern Darling Downs, Queensland. I. Effect of surface conditions on soil movement within contour bay catchments. *Australian Journal of Soil Research, 24*, 135-158.

Goos, R. J., Westfall, D. G., Ludwick, A. E. & Goris, J. E. (1982). Grain protein content as an indicator of N sufficiency for winter wheat. *Agronomy Journal, 74*, 130-133.

Green, R. E., Cornell, S. J., Scharlemann, J. P. W. & Balmford, A. (2005). Farming and the fate of wild nature. *Science, 307*, 550-555.

Hayman, P. T., Freebairn, D. M. & Huda, A. K. S. (1996). Opportunity cropping on the Liverpool Plains: a comparison of risk assessment by farmers and simulation models. In: *Proceedings of the 8th Australian Agronomy Conference* (293-296). Toowoomba, Australia: The Australian Society of Agronomy, Inc.

Hazelton, P. & Murphy, B. (2007). Interpreting Soil Test Results: *What do all the numbers mean?* CSIRO Australia.

Henderson, C. W. L. (1989). Lupin as a biological plough: evidence for, and effects on wheat growth and yield. *Australian Journal of Experimental Agriculture, 29*, 99-102.

Holt, J. A., Robertson, L. N. & Radford, B. J. (1993). Effects of tillage and stubble treatments on termite activity in two central Queensland Vertosols. *Australian Journal of Soil Research, 31*, 311-317.

Kabir, Z. (2005). Tillage or no-tillage: impact on mycorrhizae. *Canadian Journal of Plant Science, 85*, 23-29.

Kirkegaarde, J. A., Angus, J. F., Gardner, P. A. & Müller, W. (1994). Reduced growth and yield of wheat with conservation cropping. I. Field Studies in the first year of the cropping phase. *Australian Journal of Agricultural Research, 45*, 511-528.

Kuskie, J., Lawrence, D., Agius, P., Buck, S., Collins, R., Conway, M., Cox, H., Grundy, T., Lambert, G., McCosker, K., Mackay, I., Osten, V., Reid, D. & Spackman, G. (2003).

Measuring the impact of the Central Queensland Sustainable Farming Systems Project. In: B. Robinson & C. King (Eds.), *Proceedings of the 1st Australian Farming Systems Conference. Toowoomba,* Australia. Available from: http://afsa.asn.au

Li, H. W., Gao, H. W., Wu, H. D., Li, W. Y., Wang, X. Y. & He, J. (2007). Effects of 15 years of conservation tillage on soil structure and productivity of wheat cultivation in northern China. *Australian Journal of Soil Research, 45*, 344-350.

McGarry, D., Bridge, B. J. & Radford, B. J. (2000), Contrasting soil physical properties after zero and traditional tillage of an alluvial soil in the semi-arid tropics. *Soil and Tillage Research, 53*, 105-115.

Marley, J. M. & Littler, J. W. (1989). Winter cereal production on the Darling Downs – an 11 year study of fallowing practices. *Australian Journal of Experimental Agriculture, 29*, 807-827.

Perry, M. W. (1987). Water use efficiency of non-irrigated field crops. In: T.G. Reeves (Ed.), *Proceedings of the Fourth Australian Agronomy Conference* (83-89). Parkville, Victoria: The Australian Society of Agronomy, Inc.

Pillai, U. P. & McGarry, D. (1999). Structure repair of a compacted Vertisol with wet-dry cycles and crops. *Soil Science Society of America Journal, 63*, 201-210.

Radford, B. J., Gibson, G., Nielsen, R. G. H., Butler, D. G., Smith, G. D. & Orange, D. N. (1992). Fallowing practices, soil water storage, plant-available soil nitrogen accumulation and wheat performance in South West Queensland. *Soil and Tillage Research, 22*, 73-93.

Radford, B. J., Key, A. J., Robertson, L. N. & Thomas, G. A. (1995). Conservation tillage increases soil water storage, soil animal populations, grain yield, and response to fertiliser in the semi-arid subtropics. *Australian Journal of Experimental Agriculture, 35*, 223-232.

Radford, B. J., Yule, D. F., McGarry, D. & Playford, C. (2001a). Crop responses to applied soil compaction and to compaction repair treatments. *Soil and Tillage Research, 61*, 157-166.

Radford, B. J., Wilson-Rummenie, A. C., Simpson, G. B., Bell, K. L. & Ferguson, M. A. (2001b). Compacted soil affects soil macrofauna populations in a semi-arid environment in central Queensland. *Soil Biology and Biochemistry, 33*, 1869-1872.

Radford, B. J., Yule, D. F., McGarry, D. & Playford, C. (2007). Amelioration of soil compaction can take 5 years on a Vertisol under no-till in the semi-arid subtropics. *Soil and Tillage Research, 97*, 249-255.

Rawson, J. E., Bath, S. J. & Hazard, W. H. L. (1981). Effects of chemical and mechanical management of fallow land on soil moisture accumulation and yield of wheat. *Queensland Journal of Agricultural and Animal Sciences, 38*, 195-202.

Rees, R. G. & Platz, G. J. (1979). The occurrence and control of yellow spot of wheat in north-eastern Australia. *Australian Journal of Experimental Agriculture and Animal Husbandry, 19*, 369-372.

Russell, J. S. (1963). Nitrogen content of wheat grain as an indication of potential yield response to nitrogen fertilizer. *Australian Journal of Experimental Agriculture and Animal Husbandry, 3*, 319-325.

Roget, D. K., Neate, S. M. & Rovira, A. D. (1996). Effect of sowing point design and tillage practice on the incidence of rhizoctonia root rot, take-all and cereal cyst nematode in wheat and barley. *Australian Journal of Experimental Agriculture, 36*, 683-693.

Sá, J. C. M. (2004). Adubãcao Fosfatadano Sistema de Plantio Direto. In: T. Yamada, Silvia, & R.S. Abdalla (Eds.), *Sympósio sobre Fósforo na Agricultura Brasileira, Anais* (201-222). Piracicaba, São Paulo: POTAFÓS.

Sallaway, M. M., Lawson, D. & Yule, D. F. (1988). Ground cover during fallow from wheat, sorghum and sunflower stubble under three tillage practices in central Queensland. *Soil and Tillage Research, 12*, 347-364.

Shields, P. G. (1989). *Soils of the Biloela Research Station, Queensland.* Brisbane, Australia: Queensland Department of Primary Industries Research Establishments Publication No. QR89001.

Smika, D. E. & Wicks, G. A. (1968). Soil water storage during fallow in the Central Great Plains as influenced by tillage with herbicide treatments. *Proceedings of the Soil Science Society of America, 32*, 591-595.

Soil Survey Staff (1975). *Soil Taxonomy – a Basic System of Soil Classification for Making and Interpreting Soil Surveys.* Washington, DC, U.S.A: Government Printing Office, Department of Agriculture Handbook No. 436.

Standley, J., Hunter, H. M., Thomas, G. A., Blight, G. W. & Webb, A. A. (1990). Tillage and crop residue management affect Vertisol properties and grain sorghum growth over seven years in the semi-arid sub-tropics. 2. Changes in soil properties. *Soil and Tillage Research, 18*, 367-388.

Strong, W. M. & Holford, I. C. (1997). Impact of management practices on sustainability – fertilisers and manures. In: A.L. Clarke & P.J. Wylie (Eds), *Sustainable crop production in the subtropics: an Australian perspective* (214-234). Brisbane, Australia: Queensland Department of Primary Industries.

Thomas, G. A., Standley, J., Webb, A. A., Blight, G. W. & Hunter, H. M. (1990). Tillage and crop residue management affect Vertisol properties and grain sorghum growth over seven years in the semi-arid sub-tropics. 1. Crop residue and soil water during fallow periods. *Soil and Tillage Research, 17*, 181-197.

Thomas, G. A., Titmarsh, G. W., Freebairn, D. M. & Radford, B. J. (2007). No-tillage and conservation farming practices in grain growing areas of Queensland - a review of 40 years of development. *Australian Journal of Experimental Agriculture, 47*, 887-898.

Tkachuk, R. (1969). Nitrogen-to-protein conversion factors for cereals and oilseed meals. *Cereal Chemistry, 46*, 419–423.

Turpin, J.E., Bridge, B.J., Orange, D. & Thompson, J.P. (1999). Water and bromide movement in a Vertosol under four fallow management systems. *Australian Journal of Soil Research, 37*, 75-89.

Uren, N. (1991). The management of soil organic matter for sustainable agriculture. *Agricultural Science, 4*, 45-48.

Unger, P. W. (1977). Tillage effects on winter wheat production where the irrigated and dryland crops are alternated. *Agronomy, Journal, 69*, 944-950.

Walkley A. & Black L. A. (1934). An examination of the Dgtjareff method for determining soil organic matter, and a proposed modification of the chromic acid titration method. *Soil Science, 37*, 29–38.

Wildermuth, G. B., Thomas, G. A., Radford, B. J., McNamara, R. B. & Kelly, A. (1997). Crown rot and common root rot in wheat grown under different tillage and stubble treatments in southern Queensland, Australia. *Soil and Tillage Research, 44*, 211-224.

Wilson-Rummenie, A. C., Radford, B. J., Robertson, L. N., Simpson, G. B. & Bell, K. L. (1999). Reduced tillage increases population density of soil macrofauna in a semiarid environment in central Queensland. *Environmental Entomology, 28*, 163-172.

Wright, A. L., Hons, F. M., Lemon, R. G., McFarland, M. L. & Nichols, R. L. (2007). Stratification of nutrients in soil for different tillage regimes and rotations. *Soil and Tillage Research, 96*, 19-27.

Yule, D. F. (1998). Controlled traffic farming – the future. In: J. N. Tullberg & D. F. Yule (Eds.), *Proceedings of the Second National Controlled Traffic Conference* (6-12). Gatton, Australia.

In: No-Till Farming
Editor: Earl T. Nardali

ISBN: 978-1-60741-402-5

Chapter 3

IDENTIFICATION AND ASSESSMENT OF INDIGENOUS SOIL EROSION CONTROL MEASURES IN THE USAMBARA MOUNTAINS, TANZANIA

***H. B. Msita*[a], *D. N. Kimaro*[a*], *J. Deckers*[b] and *J. Poesen*[c]**

[a]Department of Agricultural Engineering and Land Planning, Sokoine University of Agriculture, P.O. Box 3003, Morogoro, Tanzania

[b]Division of Soil and Water Management, Geo-Institute, Katholieke Universiteit Leuven, Celestijnenlaan 200 E, B-3001 Heverlee, Belgium

[c]Physical and Regional Geography Research Group, Geo-Institute, Katholieke Universiteit Leuven, Celestijnenlaan 200 E, B-3001 Heverlee, Belgium

ABSTRACT

This study was conducted to identify indigenous soil erosion control measures in the Usambara Mountains, Lushoto, Tanzania. The aim was to assess the effectiveness of the indigenous soil erosion control measures in the area and their corresponding problems according to farmers' perception. The study was expected to generate new knowledge that will contribute to the efforts endeavours to link the indigenous knowledge with the recently introduced soil erosion control measures for effective soil conservation practices that are largely acceptable to the farming community. Data collection was carried out through interviews with key informants and elders in the study area, focused group discussions with community representatives, in-depth household interviews using a structured questionnaire and participant's observation during transect walks. Descriptive statistical analysis was employed in exploring the data for distribution of responses, central tendencies and dispersion. Cross tabulation and multiple response analyses was also performed to ascertain responses and percentages.The study shows that 37% of the farmer's response identified *Miraba* (Rectangular Bound Grass Strips) that do not follow contours as a unique indigenous soil erosion control practice in the Usambara Mountains. Other indigenous

* Corresponding Author: E-mail: didas@suanet.ac.tz, didas_kimaro@yahoo.com, Tel: +255 23 260 4216 (Off), +255 754 575 526 (mobile), Fax +255 23 260 1206

technologies identified include use of *Tughutu* (*Veronia subligera*) in the border of field partitions or fields (14%), and *mboji* (band of green manure) (13%). In the study area, the majority of the respondents mentioned *Miraba* (31%) as the most effective type of erosion control practice followed by contour grass strip (20%) and terraces (2%). It was observed that when Soil and Water Conservation agencies introduced contours with grass strip as a soil erosion control measure in the studied villages, they could not make distinction between *Miraba* (an indigenous technology (IT)) practiced by the farmers on one hand and grass strip on contours on the other hand as there was no critical effort made to include farmers IT in the technology development. According to elder farmers, *Miraba* were initially established to reduce the speed of the river floods runoff in the valley bottoms. However, through evolution the technology was transferred and applied on the sloping landscape where interrill, rill and gully erosion is dominant. During transect walks, severe soil erosion and gullies were observed. The interviewed key informants mentioned these erosion processes as one of the problems associated with improper terrace construction, while accelerated runoff was mentioned as the problem associated with poor implementation of the *Miraba* technique on the slopping lands. It is therefore concluded that there is a need to differentiate farmers' perception and objectives on indigenous knowledge on one hand and on the introduced technologies on the other hand in order to develop a holistic approach that include farmers IT in the overall technological development. The study underscores the importance of further research to evaluate the effectiveness of the *Miraba* technique and to improve on its performance through linkage with the recently introduced soil erosion control practices. Training on the use of introduced technologies to improve the effectiveness of indigenous technologies is strongly recommended.

Keywords: Indigenous knowledge, erosion control measures, farmers' perception, Miraba, Terraces, Usambara Mountains, Tanzania

1.0. INTRODUCTION

Indigenous soil and water conservation technologies in Africa have been recognised and documented over time (Reij *et al.,* 1996). A study conducted in Tahoua, Niger (Hassan, 1996) showed that improved traditional planting pits (*tassa*) were widely adopted by farmers as an indigenous soil erosion control technique attributed to reduced soil loss, improved soil moisture and fertility and increased millet yield. Other reported studies on indigenous technologies (IT) in Africa include for example *Zaï* on the Dogon Plateau in Mali (Wedum *et al.*, 1996) and in the Yatenga, Burkina Faso (Ouedraogo and Kaboré, 1996) where the importance of *Zaï* in the rehabilitation of abandoned lands, reduced soil erosion and improved crop yields have been demonstrated. In Ethiopian Highlands (Alemayehu, 1996) it is reported that traditional ditches are widely practiced by farmers cultivating crops on steep and infertile slopes. In this area farmers are reported to practice traditional ditches to control soil erosion and water logging regardless of other soil and water conservation methods introduced in the area. From these studies though seem that indigenous soil and water conservation technologies are still widely adopted in many areas of Africa; up scaling of such technologies from one region to another is a challenge since these practises are site specific and culturally embedded as it has also been observed in many parts of East Africa including Tanzania (Warren, 1991; Temu and Bisanda, 1996).

In Tanzania, indigenous soil and water conservation technologies practiced by communities with diverse cultural origin have been reported and documented (Floor *et al.*, 1990; Mbegu, 1996; Temu and Bisanda, 1996). In the Matengo Highlands, in Mbinga District, Tanzania *Matengo* pits (*ngolo/ingolu* system) is an indigenous technology extensively practiced to control eroding run-off water from the steep slope and hillsides (Temu and Basida, 1996; JICA, 1998). The *ngolo* system are constructed by grass laid out in strips, forming a square of rectangular, with sides varying from 100 to 200 cm grids all over the area earmarked for cultivation. The grass is then covered with topsoil, dug from the centre of the grid and spread on top of the grass leaving pits (*ingolu*) in the middle. The pits varied in depth from 15 to 60cm, and from a distance they resemble a honeycomb or chessboard. The problem associated with this technology is that construction of *ngolo* is labour intensive when compared with ridges and bench terraces (JICA, 1998) though farmers practicing this technology in the area don't feel this due to the fact that it is culturally emended.

Other reported indigenous technologies in Tanzania include *Fipa* mounds in Sumbawanga District, and *Vinyungu* in Njombe district (Mbegu, 1996), *kibanja* system in Bukoba (Floor *et al.,* 1990; Rugalema *et al.*, 1994; Baijukya, *et al.*, 2004), and the *Kihamba* system on the slopes of Mountain Kilimanjaro (O'Kting'ati *et al.*, 1984; Ikegami, 1994).

Fipa mounds are cultivation system whereby tall grass known as *kasanza*, mixed with bushes or leaves of acacia trees are slashed at the beginning of the rainy season and put into oval-shaped piles and set on fire. Later on, mounds of 90 cm high and wide are constructed in rows spaced about 30cm apart by pilling turf (soil turf) into mounds with the attached grass facing inwards covering both the ash and semi-decomposed old grass (Mbegu, 1996). This technology is practiced to improve soil fertility from compost of green manure and leguminous plants which fix nitrogen. The arrangement of mounds in rows on slopes is also important in controlling water run-off and conserving both soil fertility and moisture.

Vinyungu on the other hand are practised in valley bottoms and in uplands both during dry and wet season cultivation (Lema, 1996). A ditch or fallow is used to separate *vinyungu* from one another with the purpose to improve drainage during wet season. The burning of green vegetation is done to enhance soil fertility in particular addition of potash, while unburned green vegetation supply the soil with humus (organic matter).

The *kibanja* system extensively practiced in many areas in western part of Lake Victoria is characterised by a permanent mulch layer on the soil surface, consisting of non-edible parts of banana, including the fruit stalk and banana peels. In addition to this composition, farmyard manure mixed with grasses from *rweya* (less fertile land) and all other household refuses are also mixed within the *Kibanja*. Mulching with grasses from *rweya* is a common practice, especially near coffee trees. The result is an increase in the fertility status of the *kibanja* at the expense of the fertility status of the *rweya*. Due to the permanent ground cover, extensive weeding is not necessary and the mulch helps to conserve moisture (Rugalema *et al.*, 1994; Baijukya, *et al.*, 2004).

The *Kihamba* system is the *Chagga* tribe practice on the slopes of Mountain Kilimanjaro whereby intensive land use comprising forestry (first storey), banana (Second storey) coffee (Third storey), followed by annuals as intercrop and livestock in a sophisticated homestead farm is exercised (Ikegami, 1994; O'Kting'ati *et al.,* 1984). The *kihamba* system utilizes solar energy efficiently, where banana and all other tall trees provide shelter for coffee trees, and it help to conserve moisture and mild climate for good Arabica coffee yield. The system though is labour extensive it uses land intensively which sustains multi-cropping system, combining annuals and biennials with perennial herbs and harbour plants that enable farmers to produce the necessary living materials with least labour inputs. Despite the importance of the *kihamba* system there also a number of associated problems which include inadequate soil erosion control and heavy nutrient grass feeder inherent from Guatemala grass used as animal fodder.

In view of the examples discussed above it is obviously observed that indigenous soil and water conservation efforts in Africa and Tanzania in particular are extensively implemented by communities on specific locations where cultural values are preserved. However, interventions to improve the effectiveness and efficiency of these technologies and the rate of adoption have been minimal as it has been observed in many case studies in the Eastern Arc Mountains which include Usamabara Mountains (Watson, 1972; Temple, 1972; Conte, 1999, Sokoni and Shechambo, 2005)

Soil and water conservation in the Usambara Mountains, Tanzania has a long history (Liversage, 1944; Semgalawe, 1998; Conte, 2004). According to Watson (1972) Usambara Mountains was earmarked by the British Colonial Administration as one of the focal areas in Tanzania for implementation of projects related to increased agricultural production and soil and water conservation. According to the author, several attempts were made towards soil and water conservation in the area including closure to cultivation of steep slopes, contour ridging, and protection of stream banks and planting of grass leys for fodder and stall-feeding of stock. However, these efforts were negatively received by the local community due to the approaches that were used (Temple, 1972; Conte, 1999).

As results of the interventions mentioned above several post-independence soil and water conservation measures were introduced in the area attempting to address the past shortcomings. In addressing this situation both Catchment and participatory approaches were adopted (Theis and Grady, 1991; Pretty *et al.*, 1995; Kiara *et al.*, 1999; Kizuguto and Shelukindo, 2003; Kirway *et al.,* 2003). Some of the technologies introduced in the area following these approaches included bench terraces, strips of Napier grass (*Pennisetum purpureum*) or Guatemala grass (*Tripsacum laxum*) and different forms of agroforestry under Soil Erosion Control and Agroforestry Programme (SECAP) (Shelukindo, 1995), traditional drains and bench terrace under Traditional Irrigation Programme (TIP) (Sokoni and Shechambo, 2005), and multipurpose trees, crop strips along the contours, bench terraces, *fanya juu* infiltration ditches and cut-off drains by African Highland Ecoregional Programme (AHI) (AHI, 2000).

In addition to the above efforts several studies were also conducted in the Eastern Arc Mountains, to quantify the problems of soil erosion (FAO, 1999, Kimaro *et al.,* 2005; Vigiak *et al.*, 2006; Kimaro *et al.,* 2008), and to identify reasons for the occurrence of different forms of soil erosion (Kimaro, 2003; Tenge *et al.*, 2005) in order to come up with appropriate strategies for soil and water conservation in the area. For example in the Usambara Mountains annual soil loss amounting to 100 t ha^{-1} were reported, associated with yield reduction of most food crops to as low as 0.5 t ha^{-1} (Pfeiffer, 1990; Shelukindo and Kilasi, 1993; Lyamchai *et al.*, 1998). Wickama and Mwihomeke, 2006, conducted studies in the area to evaluate the role of IK in combating soil infertility and poverty reduction. Their results showed that there were better yield (0.8 t ha^{-1}) when farmers used the indigenous shrubs in fertilizing the soils compared to 0.5 t ha^{-1}. Despite of all these efforts made by the different programmes, soil erosion is still alarming whereby if quick measures to arrest the situation are not taken, the area will loose most of its fertile land.

The shortcomings reported above are quite often associated with mistakes committed when formulating policies for conserving and restoring the ecosystems of an area whereby local people's knowledge and skills on conservation techniques are ignored. But there is no doubt, that local community and in particular farmers around the ecosystems possess more valuable indigenous wisdom to foster the natural resource management in an undoubtedly effective ways (Warren, 1989). The Tanzanian government for many years has been disseminating technologies for soil and water conservation in the Eastern Arc Mountains including Usambara Mountains which is representative for larger mountainous areas in East Africa (Kimaro *et al.,* 2008). Surprisingly, the rate of adoption of soil and water conservation technologies in these areas is very minimal (Tenge *et al.*, 2005).

Several attempts have been made to assess the factors that have been affecting the rate of adoption of soil erosion control measures at various spatial scales (Tenge, 2005; Woldeamlak, 2007). Studies conducted in the Usambara Mountains to underscore the effectiveness and farmers' preferences of soil and water conservation techniques (Tenge, 2005) indicate that farmer's knowledge, and social economic situations are the most important factors that determine the effectiveness and farmers preference towards soil and water conservation measures. Furthermore, other studies that have been carried out to quantify soil loss reduction and the economic benefit obtained from different soil and water conservation measures show that farmers could make selection of soil and water conservation measures that are feasible under their biophysical and economic condition if they are involved in planning and informed about the cost and benefits (Tenge *et al.*, 2007).

Despite of all above, still many questions on the adoption rate of soil and water conservation technologies have remained answered. There are many assumptions that have been put forward; may be the technologies are very expensive, or are not relevant to the farmers (Tenge *at el.*, 2005). Whatever the reasons, many observations indicated that farmers have often been left out in technology generation and development process (Warren, 1989; Sokoni and Shechambo, 2005, Vigiak *et al.*, 2005). For example, Temple (1972) reported that the failure on the implementation of soil and water conservation in the Uluguru Mountains, Tanzania was partly due to disorganisation of the traditional cultivation patterns and institutions whereby farmers' indigenous knowledge and skills were largely ignored during the implementation of the conservation policies. Similar observations were also reported by Watson (1972) while working in Mlalo Basin Rehabilitation Scheme in the Usambara Mountains, Tanzania.

The alarming soil erosion in the mountainous areas of East Africa has raised a lot of ecological, environmental and economic concerns. Efforts to arrest soil erosion in these areas have progressed very slowly partly due to the fact that very little has been done to assess the contribution of indigenous knowledge on soil erosion control in arable lands, especially in mountainous areas where soil erosion processes and agricultural and other social economic activities are complex. This study was carried out to identify indigenous soil erosion control measures in the Usambara Mountains, Lushoto, Tanzania in order to assess their effectiveness and the associated problems according to farmers' perception. The study was conducted to generate new knowledge that will contribute to the efforts endeavours to link the indigenous knowledge with the recently introduced soil erosion control measures for effective soil conservation practices that are largely acceptable to the farming community in the mountainous areas of East Africa and other similar areas.

2.0. Description of the Study Area

Location

The study was conducted in two villages in the Usambara Mountains with two contrasting climatic conditions. The studied villages include Mwangoi (dry warm zone) and Migambo (humid cold zone) which are located in Mlalo and Lushoto wards respectively in Lushoto District, Tanzania (Figure 1). The approximate geographical location of the studied villages range from 417084 to 431919 E and 9471148 to 9496007 N UTM coordinates. Mwangoi Village lies approximately at an altitude of 1,430 m asl and Migambo Village at 1,651 m asl. Usambara Mountains is one of the highly populated areas in Tanzania. According to the URT (2002), the population in this area is estimated at 419,970 people and the annual growth rate is about 3%, giving a population density greater than 100 persons km^{-2}.

Climate

The Usambara Mountains varies in terms of topography and these aspects have resulted into several microclimates and soil complexes (Tenge, 2005; Conte, 1996). Generally, the area has good climatic conditions that have attracted not only farming communities but also tourists, and provide agricultural products to the population within and outside the highlands. The area has a bimodal rainfall pattern with annual rainfall ranging from 800 to 1700 mm (Tenge, 2005, Mowo *et al.* 2002).

The long rainy season is from March to May (*masika*) and the short rains (*vuli*) last from October to December. The average daily temperature ranges from 18°C to 23°C with a maximum in March and minimum in July (Wickama and Mowo, 2001).

Hydrology

The Usambara Mountains like all the other Eastern Arc Mountains is drained with several rivers and streams including river Umba, Sine, Lwengea and Zigi. The water from these rivers supports various commercial services such as the countries key hydropower generation facilities which contribute to the total capacity of more than 380 MW hydropower stations in Tanzania constituting more than 70 per cent of the total national capacity, irrigated agriculture, tourism and fisheries (Mowo *et al.*, 2002, Sokoni and Shechambo, 2005).

Land Tenure and Land Use

Different types of land tenure system exist in the study area. They include inheritance, purchase, borrowing, renting and public land. Majority of households own land under the inheritance category whereby the head of the household is the owner of the land (Lyamchai *et al.*, 1998). Generally, sixty percent of the arable land is allocated to Cash crops. Land allocation to annual crops varies with season depending on economic importance and food preference.

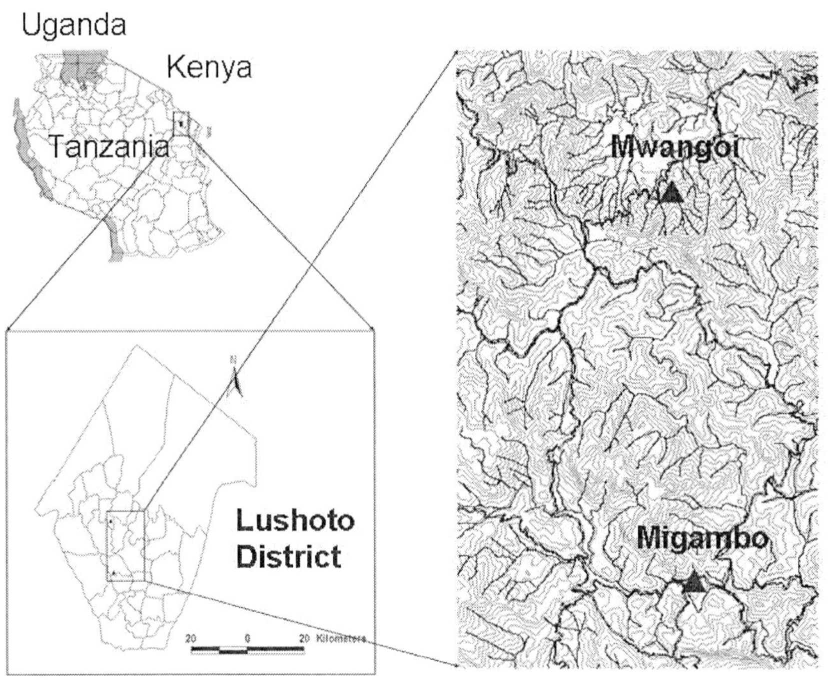

Figure 1. Location of the study area

Farming Systems and Land Use

The main economic activity in the west Usambara highland is mainly determined by the major farming systems and land uses. Agriculture is the dominant land use on which more than 90% of the population depends on it (Shelukindo and Kilasi, 1993; Lyamchai *et al.*, 1998). The major crop production includes horticultural and cereal crops and dairy farming. The area experiences stress in terms of decline in farm size and crop production due to population pressure and land degradation. To a great extent the study area is characterized by a mixed farming system, whereby farmers are involved in rainfed agriculture, traditional irrigation in valley bottoms, livestock keeping and off-farm activities. Rainfed agriculture is the most important land use followed by irrigated agriculture, livestock keeping and off farm activities including pet cash and carpentry. The major cash crops are vegetables mainly grown on valley bottoms (Figure 2), coffee, and fruits while banana, maize and beans are major food crops (Tenge, 2005). The dominant farming systems are coffee-banana intercropped with different tree species, maize-beans intercrop and patches of round and sweet potatoes, tomatoes, sugarcane or cassava. On average a household has 1.4 ha of land for rainfed agriculture. Soil erosion is one of the major constraints to agricultural production in the area such that more than one third of the catchment areas are exposed to high and very high soil erosion risk (Meliyo *et al.*, 2002).

Figure 2. Vegetable production in the valley bottoms (*vitivo*) with traditional drains

3.0. Data Collection

Informal Survey

Transect walks were held in each village (Mwangoi and Migambo villages) guided by the respective key informants who were also asked to give their opinions regarding soil erosion control measures in their area. In the field, the following checklists were used to guide the discussion:

1. If farmers were aware of existing soil erosion control measures
2. Was there any practiced indigenous soil erosion control measures?
3. What are the existing types of soil erosion control measures?
4. What is the status of soil erosion control measures practiced in the field?
5. What are the most preferred existing soil erosion control measures?
6. General success and problems associated with implementation of soil erosion control measures

In total twelve hamlets were visited in Mwangoi village which include Moa, Isoka, Kigonoroi, Amboni, Kwemiyangu, Mwangoi, Kipundili, Mabilioni, Kwediwa, Shita, Bomo and Vuga. In addition 29 farmers who have their fields in the nearby villages such as Majulai and Dule were also visited. As for Migambo Village, interviews were done in eight hamlets which include Migambo, Kwempasa, Mkuzi, Wanilo, Mhande, Kwesimu, Kwaboli and Kweshiu. In addition to hamlets visits, farmers were also interviewed in their field in which 65 farm fields were visited. The large number of fields visited in Migambo is attributed to the fact that most of the fields are located around the residence in contrast to Mwangoi village.

Formal Survey

A semi-structured questionnaire was prepared for gathering data and information on the farmers' situation regarding the effectiveness of practices in soil erosion control, and reasons for farmers' selection, perception and preference on a certain type of soil erosion control measure. The questionnaire was administered to gather data on percentage of farmers responding to the most effective cultural practices (indigenous and recently introduced practices on soil erosion control) and multiple comparisons on the effective practices. The questionnaire also facilitated to identify the indigenous practiced soil erosion control measures and the farmers preferences (Moser and Kalton 1997, Doyle, 2007).

A stratified random sampling according to gender was used during the interview in order to obtain the general characteristics of the farmers (Figure 3). The following four groups were considered: young and elders and male and female farmers. In Mwangoi the interviewed group comprised 50% male and 50% female while 60% were elders and young farmers were 40%. In Migambo, 73% of the interviewed groups were male and 27% female while elder farmers were 68% and 32% were young farmers. The groups were considered based on the type of crops they grow, cultural practices and social economic power (Tenge *et al.*, 2005). In most cases young farmers grow vegetables for selling in the nearby towns and cities, while most of the elders and female farmers cultivate bananas, maize and beans for home consumption and food security.

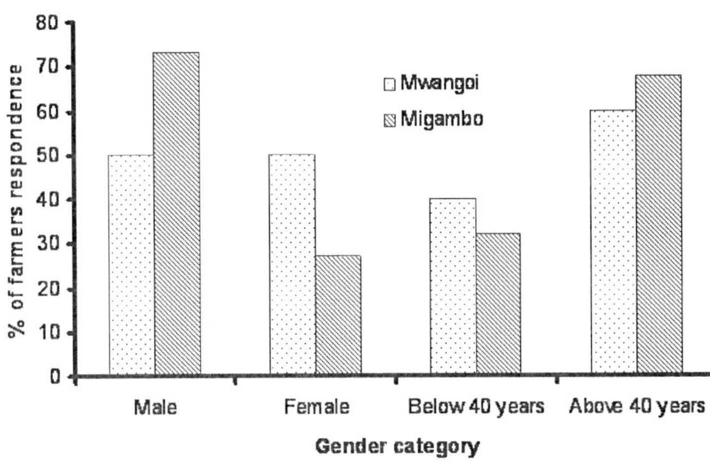

Figure 3. Percentage farmers' response according to gender

Data analysis

Descriptive statistical analysis was employed in exploring the data for distribution of responses, central tendencies and dispersion. Cross tabulation and multiple response analyses was also performed to ascertain responses and percentages

4.0. Characteristics of the Farmers in the Studied Villages

Production Characteristics

Production activities by farmers in the studied villages are presented in Figure 4. The results showed that in the two villages agriculture is dominantly crop based. Crop producers account for about 97% and 88% of all agricultural activities in Mwangoi and Migambo Villages respectively. On the other hand, there more livestock-keepers in Migambo (5%) when compared to Mwangoi (0%) (Figure 4). The study revealed also that in Migambo more people depend on pet cash (7%) than in Mwangoi village (3%). Dependence on agriculture in both villages is attributed to favourable climatic conditions i.e. relatively adequate rainfall for most of the crops grown in the area. However, in Migambo village rainfall reliability is more certain than Mwangoi and hence an incentive to engage in different farming activities such as livestock keeping due to readily availability of animal feeds (pasture), and fruit production. Furthermore, the results showed that in Migambo village there are more pet cash activities due to its proximity to Lushoto town and nearby tourist hotels (Mowo *et al.*, 2002).

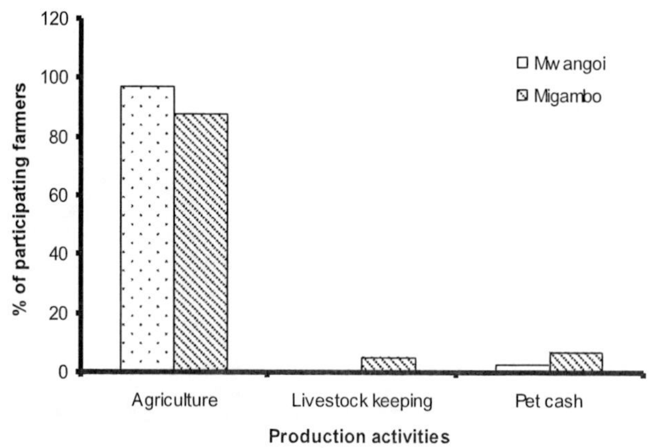

Figure 4. Farmers production activities

Farmers' Production Institutions

In both villages farmers were organized in different community production groups based on agricultural activities such as; horticultural, poultry and livestock keeping, apiculture and microfinance institutions including Savings and Credit Co-operative Societies (SACCOS) (URT, 2000, Rubambey 2005). The study revealed that about 25% of the respondents were associated with horticulture farming groups in Mwangoi while in Migambo Village only 7% of the interviewed farmers were organized in horticultural groups (Figure 5). This could be explained by weather conditions whereby temperatures in Mwangoi favour more production of vegetable crops in contrast to a cooler climate in Migambo. In Migambo village the climatic conditions are more favourable for production of temperate fruit crops which is not the case for Mwangoi village.

Sensitivity of vegetable handling and processing for market require extra efforts due to their perishable nature. Farmers in Mwangoi have organised themselves into groups in order to cope with market forces associated with vegetable crop production including value addition so as to obtain favourable benefits. The study also showed that in Migambo village livestock keeping including dairy and poultry is a growing farming activity. Due to the fact that these enterprises require substantial investment, farmers in Migambo Village are more organised in microfinance institutions (SACCOS) as expressed by the 33% of the interviewed respondents than Mwangoi (3%). In addition to farming and microfinance institutions, it is also revealed from this study that farmers in both Mwangoi and Migambo village are also organised into natural resource management (NRM) groups such apiculture as a mechanism towards rehabilitation of degraded forest.

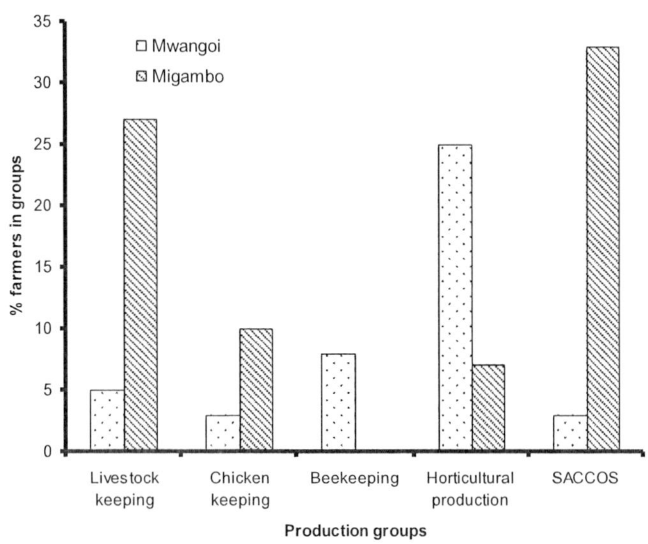

Figure 5. Farmers production institutional groups

This was facilitated by the Government intervention through Participatory Forest Management (PFM) policy (Meghji, 2003, MNRT 2006) and a research named "Promoting suatainable natural resources management through effective governance and farmer-market linkage" a regional collaborative research project implemented by Uganda, Kenya and Tanzania through Associatiaion for Strenthening Agricultural Research in East and Central Africa (ASARECA) (Semalulu *et al.*, 2005).

Farmers' Level of Education

Figure 6 present the level of farmers at various education categories in the studied villages. The results showed that most of the farmers had primary education in both Mwangoi (85%) and Migambo (65%), while in both villages only few had secondary education (7%).

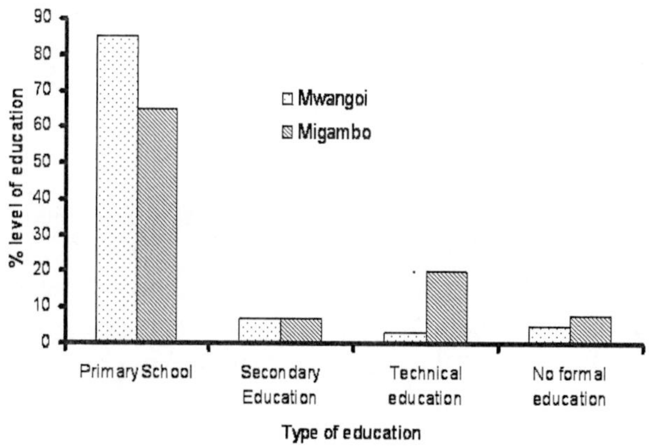

Figure 6. Percentage household Level of education

Table 1. Number of field plots owned by family

Number of field plots	Response (%)
Mwangoi	
1-3	17
3-6	67
6-9	13
9-12	0
12-15	3
Migambo	
1-3	33
3-6	52
6-9	10
9-12	2
12-15	3

Due to the influence by the Roman Catholic Church in Vocational Education and Training (VETA) located in Migambo village, about (20%) of the respondents in Migambo village had training in vocational skills. Similar observations were made by Sokoni and Shechambo, (2005) in west Usambara when evaluating changes in the upland irrigation system and its implications for Rural Poverty Alleviation.

Farm Plot Size and Distribution Per Household

Generally farm sizes in both villages are small ranging from 0.4 ha to 2.4 ha per household (Table 1 and Figure 7). Field observation showed that in Mwangoi village farm sizes are more fragmented than in Migambo village attributed to relatively high population density (Tenge, 2005).

Labour Organisation

The results showed that labour requirements for farming activities in the studied villages were mainly family labour (Figure 8). Further, it was observed that family labour contribution in mwangoi was higher (77%) compared to Migambo (60%). Similar trends were also observed by Ngailo *et al.* (1998) and Lyamchai *et al.* (1998) in other villages in the Usambara Mountains.

5.0. Farmers Soil Erosion Control Measures Identified in the Study Area

Table 2 present different types of soil erosion control measures as identified by farmers in the two studied villages. In both villages most farmers were aware of the recently introduced soil erosion control technologies such as bench terraces, *fanya juu*, contouring, drainage ditches, and cut off drains.

Table 2. Farmers awareness response on type of soil erosion control

Soil erosion control technologies	Farmers' response (%)
Mwangoi	
Terraces	53
Drains	12
Contours	10
Fanyajuu terraces	4
Ditches	8
Green manure	10
Soil erosion control technologies	Farmers' response (%)
Migambo	
Terraces	80
Drains	3
Contours	7
Fanyajuu terraces	0
Farmyard manure	3
Green manure	0

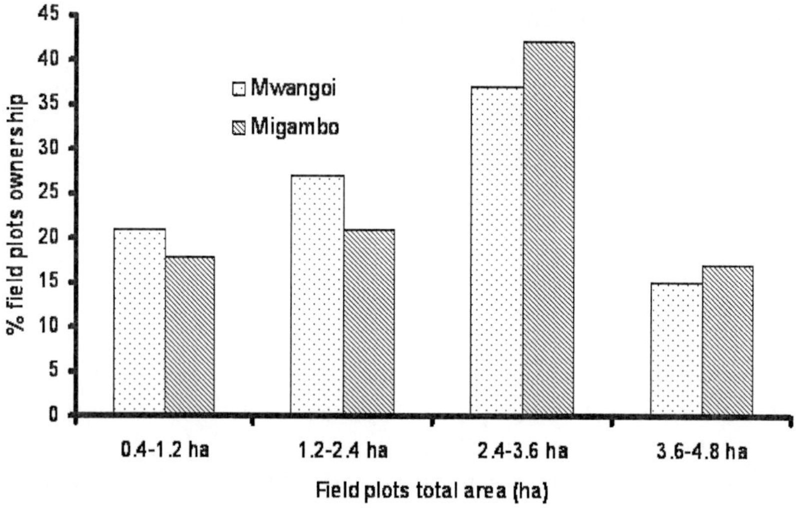

Figure 7. Total area of field plots owned by household members

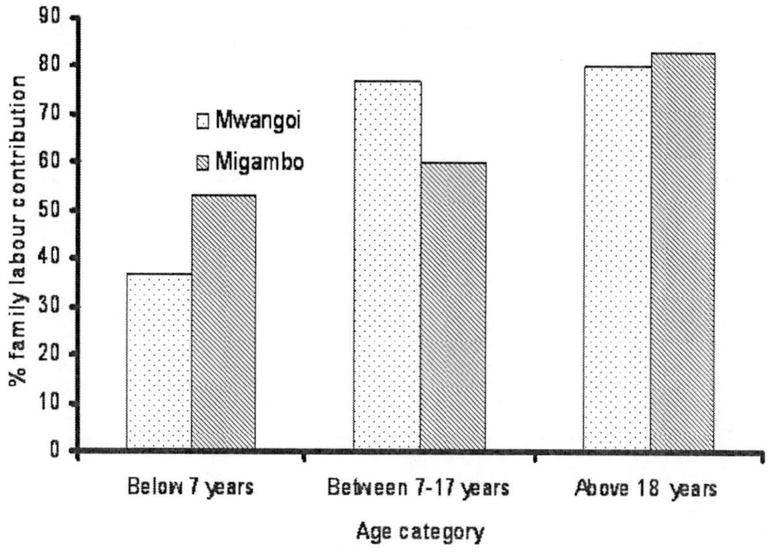

Figure 8. Family labour contribution according to age

This awareness had been enhanced by the various soil erosion control measures implemented by various programmes and institutions (Figure 9) in the area such as Soil Erosion Conservation and Agroforestry programme (SECAP) (Shelukindo, 1995). The thrust of this programme was mainly on agroforestry to conserve water and soil erosion control. Other activities implemented under this programme included planting of Napier grass (*Pennisetum purpureum*) or guatamala grass (*Tripsacum laxum*) along the contours.

Terraces were reported by farmers to be dominant soil erosion control measures on sloping land and mountain hills. As a result of the conservation practices introduced in the area, some farmers adopted implementation of terraces with application of mulching and or farmyard manure aimed to improve soil fertility and water availability to crops (Lyamchai *et al.*, 1998 and Tenge, 2005). As shown in Figure 10 about 77% and 35% in both Mwangoi and Migambo respectively applied green manure on the terraces than the other type of fertilizers. Similar observation was also made by Tenge, (2005) while evaluating the physical effectiveness of various soil erosion control measures in the East African highlands.

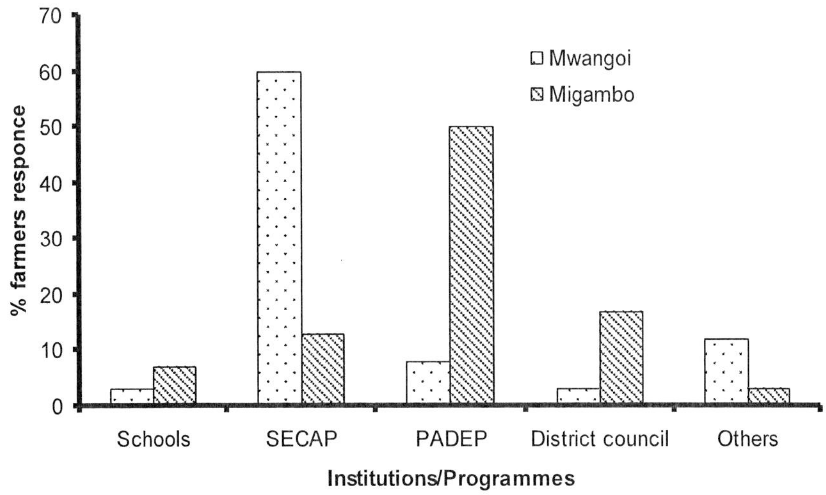

Figure 9. Response of Farmers' awareness on institutions/programmes implemented soil and water conservation technology in the studied villages

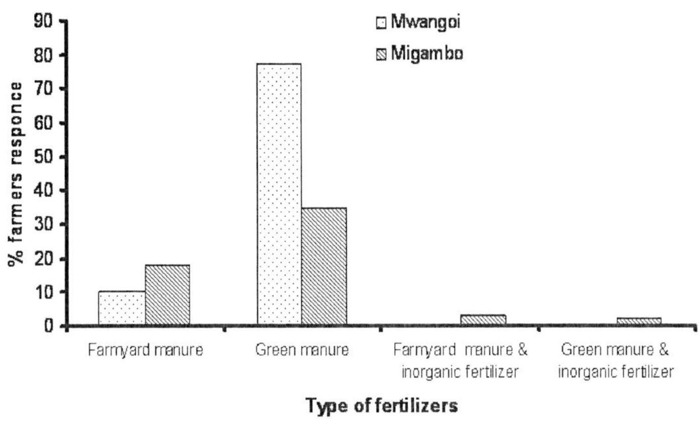

Figure 10. Percentage farmers' response on type of fertilizers used

6.0. Indigenous Soil Erosion Control Technologies Identified in the Study Area

Figure 11 present a list and farmers' response on the identification of indigenous soil erosion control technologies in the studied villages. The study shows that 37% of the farmer's response identified *Miraba* (Rectangular Bound Grass Strips) that do not follow contours as a unique indigenous soil erosion control practice in the Usambara Mountains. Other indigenous technologies identified include use of *Veronia subligera* (*Tughutu*) in the border of field partitions or fields (14%), and band of green manure (13%). *Miraba* was mainly identified as indigenous soil erosion control measures by the aged farmers (above 40 years) visited during transect walks.

In the past, before colonial rule in 1900s, farmers used to measures irregular rectangular field plot boundaries where they planted some vegetation along the enclosed plots (*miraba*) (Figure 12) to prevent soil movement caused by runoff in the valley bottoms (*vitivo*) (Figure 13) and on the sloping land, where most of the crop farming were done .

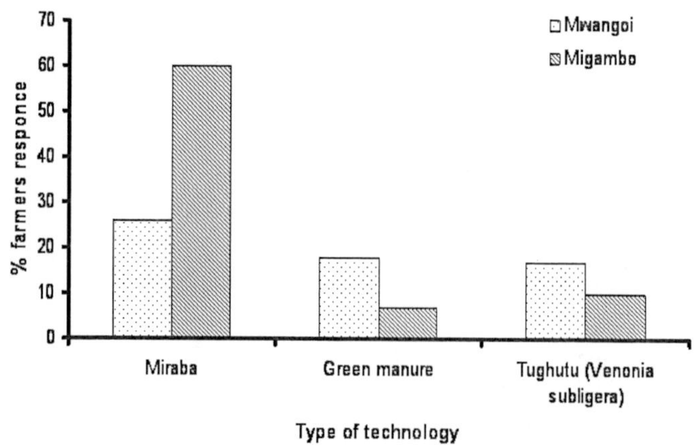

Figure 11. Percentage of farmers' response on the identification of indigenous soil erosion control technologies

Figure 12. *Miraba* technology practiced in Migambo village

Figure 13. Miraba in the valley bottom fields in Mwangoi and Migambo village

On the hand and in accordance to the interviewed farmers, *tughutu* was a soil and water conservation technology practiced in the area before recently introduced soil erosion control measures were brought in mainly by the colonial rule in the early 1940s and 50s. Before this period, farmers used to measures the strips along field plot boundaries by foot steps (not exactly on the same distance) where they planted *Veronia subligera* along the line (*Tughutu*) (Figure 14) which improved soil fertility and controlled runoff.

Figure 14. *Tughutu* technology practiced in Mwangoi village

Alongside *miraba* and *Tughutu* farmers used to produce *mboji* (green manure) (Figure 15) in flat areas whereby weeds and other crop residues were hipped and left to decompose and used as a fertilizer in the coming farming season. *Mboji* was mainly applied by farmers to enhance soil fertility whereby in the long run reduced runoff through improved soil structure. The use of *Mboji* (green manure) and *tughutu* IK was also observed by Wickama and Mwihomeke (2006) while investigating the role of IK on soil fertility improvement in the Usambara Mountains, Tanzania. It should be noted that most of the respondents identified the indigenous technologies from the recently introduced soil erosion control measures by comparing the technologies practiced during pre and post-colonial periods as most of the indigenous technologies on soil erosion control measures were taught by elders while working in the field, and instructions which were given during informal education and cultural rituals.

Figure 15. *Mboji* technology practiced in Mwangoi village

Similar observation was made by Rajasekaran (1993) when formulating a framework for incorporating indigenous knowledge systems into agricultural research, extension, and NGOs for sustainable agricultural development in India.

In this study it was observed that farmers in the past used Indigenous Technologies (IT) holistically. However, recently there has been a tendency of farmers preferring to combine IT with farm yard manure. This practice is observed to be an influence from recently introduced soil erosion control technologies whereby farmers were taught to combine for example bench terrace, contours, and cut-off drains with farm yard manure. Hence, in the studied villages, it was observed that farmers preferred to practice *miraba* and terraces with farmyard manure (Figure 16) than other technologies. Observations from other studies attributed this to the fact that farmers prefer technologies that generate quick response and both yield and economic benefits (Warren, 1989; Samantha, 1996; Samantha, 2002; Wickama and Mwihomeke, 2006).

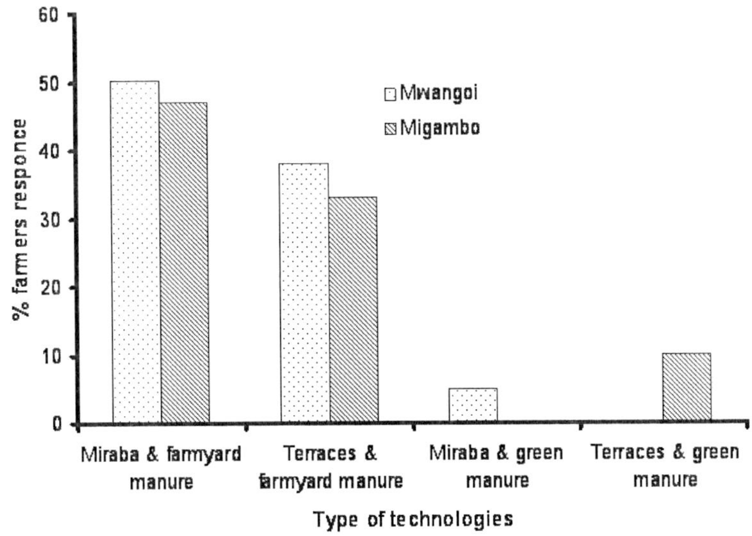

Figure 16. Farmers preference response on the application of both indigenous and recently introduced soil erosion control technologies with manure

7.0. Problems Associated with Soil Erosion Control Technologies

It was observed that when Soil and Water Conservation agencies (extension workers) introduced contours with grass strip as a soil erosion control measure in the studied villages, they could not make a distinction between *Miraba* (an IT practiced by the farmers) on one hand and grass strip on contours. However, farmers were more acquainted with *miraba* than the introduced contour technology. Tenge *et al.,* (2004) and Tenge (2005) while working in Kwalei village in the Usambara Mountains, Tanzania reported only the existence of grass strips and other types of recently introduced soil erosion control technologies but not *miraba*. During the survey it was also observed that most of the extension agents could not recognise *miraba* as a different form of indigenous soil erosion control technology from contour grass strip. However, farmers as shown in Figure 17 could make a clear distinction between *miraba* and contour grass strips. This observation has a negative implication on technological transfer and linkage between indigenous and introduced technologies. Therefore, there is a need for critical effort to merge indigenous and introduced soil erosion control technologies in the area. It is therefore concluded that there is a need to differentiate farmers' perception and objectives on indigenous knowledge on one hand and on the introduced technologies on the other hand. The study underscores the importance of further research to evaluate the effectiveness of the *Miraba* technique and to improve on its performance through linkages with the recently introduced soil erosion control practices. Training on the use of introduced technologies to improve the effectiveness of indigenous technologies is strongly recommended.

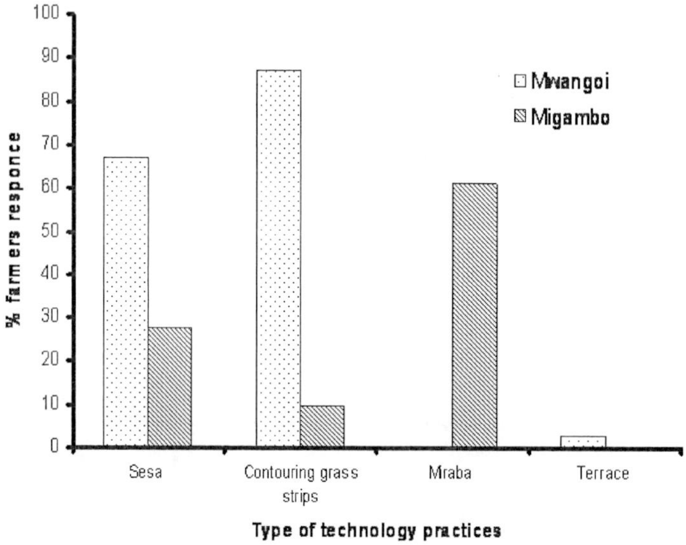

Figure 17. Farmers' response on distinction between various soil erosion control technologies

Figure 18. Gullies associated with bench terrace construction in Mwangoi village

According to elder farmers, *Miraba* technology was initially established to reduce the speed of river floods runoff in the valley bottoms. However, through evolution the technology was transferred and applied on the sloping landscape where interrill, rill and gully erosion is dominant. During transect walks, severe soil erosion and gullies were observed (Figure 18). The interviewed key informants mentioned these erosion processes as one of the problems associated with improper terrace construction, while accelerated runoff was mentioned as the problem associated with poor implementation of the *Miraba* technique on the slopping lands (Figure 19).

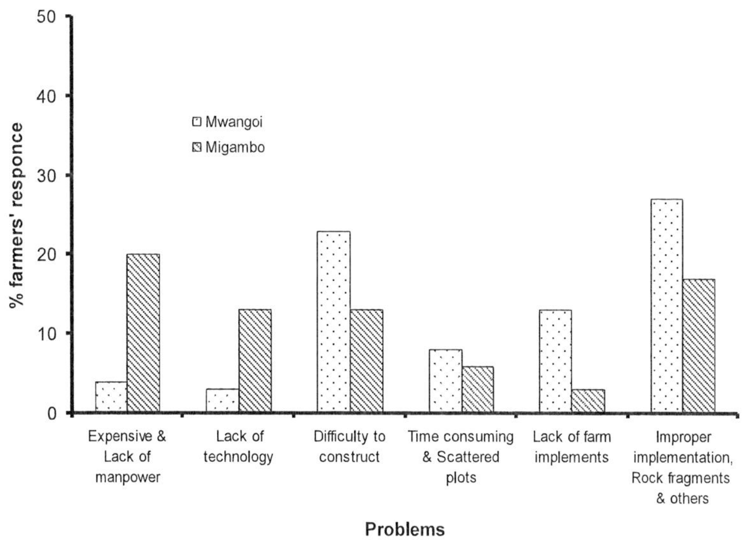

Figure 19. Farmers' response on problems associated with implementation of soil erosion control technologies

Field observation showed that when *miraba* are constructed without following contours, it is most likely rain water runoff would be blocked creating temporary ponding which eventually overrun along the *miraba* causing damage in the downward fields. However, this is an area which requires further research.

Other problems associated with implementation of soil erosion control technologies are also given in Figure 19. Inadequate technological know how, manpower and costs related to the construction of soil erosion control structures are some of the problems associated with implementation of soil and water conservation technologies reported by farmers in the area. For example, in Migambo village inadequate technological know how (13%), manpower requirements and associated costs (20%) are the reported problems associated with implementation of soil and water conservation technologies. In Mwangoi village, majority of farmers (more than 23%) reported that physical land conditions such as rock fragments and infertile soils coupled with inadequate tools as problems which have resulted into multiple difficulties in the implementation of soil and water conservation technologies. Very few farmers in both villages (<8%) have reported time constraints attributed to scattered plots as a major problem associated with the implementation of soil and water conservation technologies.

8.0. CONCLUSION

The results of this study demonstrate that farmer's in the Usamabra Mountains, Tanzania recognise *Miraba* (Rectangular Bound Grass Strips) that do not follow contours as a unique indigenous soil erosion control practice in the area. Other indigenous technologies that are important in the area according to the farmers include use of *Tughutu* (*Veronia subligera*) in the border of field partitions and *mboji* (band of green manure).

The study shows that most agents responsible for scaling up soil erosion control technologies could not recognise *miraba* as a different form of indigenous soil erosion control technology from contour grass strip. However, farmers could make a clear distinction between *miraba* and contour grass strips. It is therefore concluded that there is a need to differentiate farmers' perception and objectives on indigenous knowledge on one hand and on the introduced technologies on the other hand in order to develop a holistic approach that include farmers IT in the overall technological development.

According to farmers' perception, *Miraba* which were initially established to reduce the speed of river floods runoff in the valley bottoms are inadequate in controlling soil erosion related to accelerated runoff on the slopping lands. Insufficient technological know how, manpower and costs related to the construction of soil erosion control structures are problems associated with the implementation of indigenous soil erosion control technologies and other soil and water conservation practices reported farmers in the Usamabara Mountains.

The study underscores the importance of further research to evaluate the effectiveness of the *Miraba* technique and to improve on its performance through linkages with the recently introduced soil erosion control practices. Training on the use of introduced technologies to improve the effectiveness of indigenous technologies is strongly recommended.

ACKNOWLEDGMENT

The authors wish to express their sincere appreciation to Belgium Technical Cooperation (BTC) and the Research Project on Promoting Sustainable Natural Resource Management through Effective Governance and Farmer-Market Linkages funded by the European Union through the Association for Strengthening Agricultural Research in Eastern and Central Africa (ASARECA) for the financial assistance towards this work. Farmers in the Usamabara Mountains particularly the residence of Mwangoi and Migambo Villages are highly acknowledged.

REFERENCES

AHI (African Highlands Initiative), (2000). Annual Report. District Agricultural Office Lushoto, Tanzania.

Conte, C. (1996). Nature reorganised. Ecological History in the Plateau Forests of the Usambara Mountains. In *custodians of the Land, Ecology and culture in the History of Tanzania,* Maddox, G., Giblin, J. L. & Kimambo, I. (Eds). Mkuki na Nyota, Dar es Salaam.

Conte, C. (1999). Colonial Sciences and Ecological Change: Tanzania Mlalo Basin, 1888-1946. *Environmental History*, *4*, 220-245.

Conte, C. A. A. (2004). Highland Sanctuary: *Environmental History in Tanzania's Usambara Mountains.* African Studies Quarterly, Volume 8, Issue 2 Spring 2005. OH: Ohio University Press, 215 pp.

Doyle J. K. (2007). Introduction to survey methodology and design: Back to SySurvey Whitepapers. Chapter 10 from the book Handbook for IQP Advisors and Students, prepared by Doglas W. Woods, Professor Emeritus. http://www.sysurvey.com/tips/introduction_to_survey.htm visited on 11.11.2007.

FAO. (1999). Erosion-induced loss in soil productivity and its impacts on agricultural production and food security. Intergrated Soil Management for Sustainable Agriculture and Food Security in Southern and East Africa, Proceedings of the Expert Consultation, Harare, Zimbabwe, 8th – 12th Dec. 1997). Edited by Nabhan, H., Mermut, A. R. & Mashali, A. M. Table 4 pp 105. ftp://ftp.fao.org/agl/agll/docs/misc23.pdf , visited on 16th June 2007.

Floor, J., Kimaro, D. N., Vlot, J. E. & Van Kekem, A. J. (1990). Soil Fertility Aspects and Sustainability of the Traditional Farming System in Bukoba District, Tanzania. In Magoggo, J. P. Shayo-Ngowi, A. J. & Mnkeni, P. N. S. (eds). Proc. of the tenth Annual General Meeting of the Soil Science Society of East Africa, Dec. 1990, *Arusha Tanzania*, (PP 1-12)

Hassan, A. (1996). Improved traditional planting pits in the Tahoua Department (Niger): an example of rapid adoption by farmers; In Sustaining the soil, Indigenous Soil and Water Conservation in Africa. Edited by Reij, C., Scoones, I. & Toulmin, C. pp 56-61.

Ikegami, K. (1994). The traditional agrosilvipastoral complex sytem in the Kilimanjaro region, and its implications for the Japanese-assisted lower Moshi irrigation project. Faculty of Agriculture, Kinki University *African Study Monographs, 15(4)*, 189-209.

JICA, (1998). Integrated Agroecological Research of the Miombo Woodlans in Tanzania. Final Report. Faculty of Agriculture, Sokoine University of Agriculture and Center for African Area Studies, Kyoto University, pp 413.

Kiara, J. K., Munyikombo, L. S., Mwarasomba, L. S., Pretty, J. & Thompson, J. (1999). Impacts of the catchments approach to soil and water conservation : Experiences of the Ministry of Agriculture, Kenya. *In fertile ground. The impacts of participatory watershed management*, Hinchcliffe, F., Thompson, J., Pretty, J., Guijt, I. & Shah, P. (Eds). Intermediate Technology Publication Ltd, 103-105 Southampton Row, London WC1B 4HH, UK: 130-142.

Kimaro, D. N. (2003). Assessment of major forms of soil erosion in the Morningside catchment, Uluguru Mountains, Tanzania. PhD Thesis, Sokoine University of Agriculture.

Kimaro, D. N., Deckers, J. A. Poesen, J., Kilasara, M. & Msanya, B. M. (2005). *Short and medium term assessment of tillage erosion in the Uluguru Mountains, Tanzania*. Soil & Tillage Research. *81*, 97-108.

Kimaro, D. N., Poesen, J., Msanya, B. M. & Deckers J. A. (2008). Magnitude of soil erosion on the northern slope of the Uluguru Mountsins, Tanzania: *Interrill and rill erosion*. Catena *75*, 38-44, ISSN 0341-8162.

Kirway, N. T., Lemaa, N. M., Lyimo, S. D., Kileo, R. O., Kapange, C. & Schrader, T. (2003). *Farming systems approaches*. Training manual Volume 1. Ministry of Agriculture and Food Security, Tanzania.

Kizughuto, T. M. & Shelukindo, H. B. (2003). *The challenges facing the catchment approach in the East African highlands, the case of the Usambara mountains experiences and way forward*. SECAP discussion paper. Soil Erosion and Agro-forestry program (SECAP), Lushoto District, Tanzania.

Lema, A. (1996). Cultivating the valleys: *vinyungu* farming in Tanzania. In Sustaining the Soil, Indigenous Soil and Water Conservation in Africa. Edited by Reij, C., Scoones, I. & Toulmin, C., pp 139-144.

Liversage, V. (1944). Rural reconstruction in South Africa, E. *Afr. Agric. For. J*, *10*, 120-124

Lyamchai, C., Ndondi, R. V., Owenya, M., Ndakidemi, P. & Masawe, N. (1998). *Participatory Rural Appraisal in Kwalei catchment Lushoto, Tanzania*. In Lyimo, S. D. & Ndondi, R. V., (Eds). Selian Agricultural Research Institute, Arusha, Tanzania.

Lyamchai, C., Owenya, M., Ndakidemi, P. & Massawe, N. (1998). A Participatory Rural Appraisal in Kwalei Catchment, Lushoto District Tanzania. African Highland Ecoregional Program. Arusha: Selian Agricultural Research Institute.

Mbegu, A. C. (1996). Making the most of compost: a look at wafipa mounds in Tanzania. In *Sustaining the Soil, Indigenous Soil and Water Conservation in Africa*. Edited by Reij, C., Scoones, I. & Toulmin, C., pp 134-138.

Meghji, Z. H. (2003). Participatory Forest Management in Tanzania. Paper submitted to the XII World Forestry Congress, Quebec City Canada (0813-C1). http://www.fao.org/DOCREP/ARTICLE/WFC/XII/0813-C1.HTM visited on 06/03/2009.

Meliyo, J., Kabushemera, W. & Tenge, A. J. (2002). *Characterization and mapping soils of Kwalei subcatchment*, Lushoto district. Agricultural Research Institute, Mlingano, Tanga, Tanzania.

MNRT (Ministry of Natural Resources and Tourism). (2006). *National Forestry and Beekeeping Programme*. Ministry of Natural Resources and Tourism, Forestry and Beekeeping Division, Tanzania. http://www.nfp.co.tz/programmes.html visited on 06/03/2009.

Moser Kalton. (1997). Survey methods in social investigation. ELBS and Heinman educational books.

Mowo, J. G., Mwihomeke, S. T., Mzoo, J. B. & Msangi, T. H. (2002). Managing natural resources in the West Usambara mountains: A glimmer of hope in the horizon. http://www.mtnforum.org /resources/library/mowo02a.htm visited on 27 May 2007.

Ngailo, J. Wickama, J. M. & Nyaki A. S. (1998). Nutrient flow Analysis for Selected Farming Systems in Northern Tanzania: The case of Mbulu, Moshi rural, and Lushoto districts. Report from a FAO sponsored project.

O'Kting'ati, A., Maghembe, J. A., Fernandes E. C. M. & Weaver, G. H. (1984). *Plant species in the Kilimanjaro Agroforestry System*. The International Council for Research in Agroforestry, Agroforestry Systems 2, pp. 177-186. Martinus Nijhoff/ Dr. W. Junk Publishers, Dordrecht.

Ouedraogo, M. & Kabore. (1996). The *Zaï*: a traditional technique for the rehabilitation of degraded land in the Yatenga, Burkina Faso. In *Sustaining the Soil, Indigenous Soil and Water Conservation in Africa*. Edited by Reij, C., Scoones, I. & Toulmin, C., pp 80-84.

Pfeiffer, R. (1990). Sustainable agriculture in practice- The production potential and environmental effects of macro-contour lines in west Usambara mountains, Tanzania. PhD dissertation. Hohenheim University.

Pretty, J. N., Thompson, J. & Kiara, J. (2005). *Agricultural regeneration in Kenya: The catchments approach to soil and water conservation*. Ambio, *24*, 7-15.

Rajasekaran, B. (1993). A framework for incorporating indigenous knowledge systems into agricultural research, extension, and NGOs for sustainable agricultural development. Studies in Technology and Social Change No. 21. Ames, IA: Technology and Social Change Program, Iowa State University.

Reij, C., Scoones, I. & Toulmin, C. (1996). Sustaining the Soil Indigenous Soil and Water Conservation in Africa. Vrije Universiteit Amsterdam. ISBN 185383 372X Published in the UK by EARTHSCAN Publications Limited. pp 1-27

Rubambey, G. C. (2005). Policy, Regulatory and Supervisory Environment for Microfinance in Tanzania. Essays on Regulation and Supervisotion No 15. http://microfinancegateway.com/files/29886_file_Tanzania_Essay_Ira.pdf visited on 06/03/2009.

Rugalema, G. H., O'Kting'ati, A. & Johnsen, F. H. (1994). The home gardern agroforestry system in Bukoba district, North-West Tanzania. Farming system analysis. *Agroforestry systems* 26, 53-64 pp. Kluwer Academic Publishers. Printed in the Netherlands.

Samantha, J. (1996). Discourses on land Degradation in the Uluguru Mountains, Tanzania: Evolution and Influences. *Journal of Rural Studies, Vol. 12, No. 2* pp, 187-199.

Samantha, J. (2002). A framework for understanding on-farm environmental degradation and constraints to the adoption of soil conservation measures: case studies from highlands Tanzania and Thailand. World Development volume 30 Issue 9, pg 1607-1620.

Semalulu O., Kimaro, D.N., Ayaga, G.O., Kasenge, V., Kayanga, S.T., and Hella, J.P. (2005). Promoting suatainable natural resources management through effective governance and farmer-market linkage. ASARECA Competitive Grant System (ASARECA-CGS). Full Project Proposal. ASARECA, Entebe, Uganda.

Semgalawe, Z. M. (1998). Household adoption behaviour and agriculture sustainability in the northern mountains of Tanzania. The case of soil conservation in the North Pare and West Usambara mountains. PhD Thesis. Wageningen Agricultural University.

Shelukindo, H. & Kilasi, G. (1993). Development experience and re-orientation of soil erosion control and agro-forestry activities in West Usambara mountains, Tanzania. A paper presented at the workshop on Natural agroforesty and environment. SUA, Morogoro (12-13 October 1993).

Shelukindo, H. (1995). Technical Recommendations for soil and water conservation measures and agroforestry systems. SECAP and TIP, DALDO, Lushoto.

Sokoni, C. H. & Shechambo, T. C (2005). Changes in the Upland Irrigation System and Implications for Rural Poverty Alleviation. *A Case of the Ndiwa Irrigation System, West Usambara Mountains, Tanzania.* Research and Poverty Alleviation (REPOA). Research Report No. 05.1. ISBN 9987 417 50 7. Mkuki na Nyota Publishers LTD, P. O. BOX 4246, Dar Es Salaam, Tanzania.

Temple, P. H. (1972). Soil and Water conservation Policies in the Uluguru Mountains, Tanzania. Geografiska Annaler. Series A, *Physical Geography, Vol. 54, No., ¾*. Studies of Soil Erosion and Sedimentation in Tanzania pp 110-123.

Temu, A. E. M. & Bisanda, S. (1996). Pit cultivation in the Matengo Highlands of Tanzania. In *Sustaining the Soil, Indigenous Soil and Water Conservation in Africa*. Edited by Reij, C., Scoones, I. & Toulmin, C., pp 145-150.

Tenge, A. J. B. Okoba, O. & Sterk, G. (2007). Participatory Soil and Water Conservation Planning using a Financial Analysis tool in the West Usambara Highlands of Tanzania. *Land Degradation & Development*, *18*, 321-337.

Tenge, A. J. M. (2005). Participatory appraisal for farm level soil and water conservation planning in West Usambara highlands, Tanzania. PhD Thesis submitted, Wageningen University dissertation no. 3733. http://library.wur.nl/wda/abstracts/ab3733.html. visited on 19th June 2007.

Tenge, A. J., De Graaff, J. & Hela, J. P. (2004). Social and Economic Factors Affecting the Adoption of Soil and Water Conservation in West Usambara Highlands, Tanzania. Wiley Interscience Publisher. *Land Degradation and Development, 15*, 99-114 pp.

Tenge, A. J., De-graaff, J. & Hella, J. P. (2005). Financial efficiency of major soil and water conservation measures in West Usambara highlands, Tanzania. Applied Geography 25 (2005) 348–366, www.sciencedirect.com visited on 27th May 2007.

Theis, J. & Grady, H. M. (1991). Participatory rapid appraisal for community development. *A training manual based on experiences in Middle East and North Africa.* Save the children Federation, London, International Institute for Environment and Development, West port.

URT (United Republic of Tanzania). (2000). Report on National Micro-Finance Policy. Printed by the Government Printer Dar es Salaam, Tanzania. http://www.tanzania.go.tz/ pdf/ nationalmicrofinancepolicy.pdf visited on 06/03/2009.

URT (United Republic of Tanzania). (2002). National Census, Dar es Salaam, Tanzania.

Vigiak, O. & Emiel van Loon, Sterk, G. (2006). Modelling spatial scales of water erosion in the West Usambara Mountains of Tanzania, Geomorphology 76 (2006) 26-42. www.sciencedirect.com. Visited on 27th May 2007.

Vigiak, O., Okoba, B. O., Sterk, G. & Stroosnijder, L. (2005). Water erosion assessment using farmers' indicators in the West Usambara Mountains, Tanzania. CATENA 64 (2-3), 307-320.

Warren, D. (1991). *Using indigenous knowledge in agricultural development, Discussion paper 127*, World Bank, Washington, DC.

Warren, D. M. (1989). Indigenous knowledge systems. Implication for agriculture and International development. *The academy for Educational Development*, Inc, Washington. C.

Watson, R. J. (1972). Conservation Problems. Policies and the origins of the Mlalo Basin Rehabilitation Scheme, Usambara Mountains. Tanzania. Geografiska Annaler. Series A, *Physical Geography, Vol. 54, No. ¾, Studies of Soil Erosion and Sedimentation in Tanzania.* pp. 221-226.

Wedum, J., Doumbia, Y., Sanogo, B., Dicko, G. & Cissé, O. (1996). Rehabilitating degraded land: *Zaï* in the Djenné Circle of Mali. In *Sustaining the Soil, Indigenous Soil and Water Conservation in Africa*. Edited by Reij, C., Scoones, I. & Toulmin, C., pp 62-68.

Wickama, J. M. & Mowo, J. G. (2001). Indigenous nutrient resources in Tanzania. Managing African Soils. Issue Paper No. 21. *International Institute for Environment and Development*, London.

Wickama, J. M. & Mwihomeke, S. T. (2006). The Role of Indigenous Knowledge in Combating Soil Infertility and Poverty in the Usambara Mountains, *Tanzania. Research and Poverty Alleviation (REPOA).* Research Report No. 06.2. ISBN 9987-449-21-2. Mkuki na Nyota Publishers LTD, P. O. BOX 4246, Dar Es Salaam, Tanzania.

Woldeamlak, B. (2007). Soil and water conservation intervention with conventional technologies in northwestern highlands of Ethiopia: Acceptance and adoption by farmers, Land Use Policy 24, (2007) 404–416, www.sciencedirect.com. Visited on 20th September, 2007.

In: No-Till Farming
Editor: Earl T. Nardali

ISBN: 978-1-60741-402-5

Chapter 4

FERTILIZATION AND CROP RESIDUE MANAGEMENT IN NO-TILL CORN PRODUCTION

***B. L. Ma*[*1] *and Maria Cigales Rivero*[2]**

[1]Agriculture and Agri-Food Canada, Eastern Cereal and Oilseed Research Centre (ECORC), 960 Carling Avenue, Ottawa, Ontario, Canada K1A 0C6.

[2]Universidad de Colima, Centro Universitario de Investigacion para el Desarrollo Agropecuario (CUIDA), Colima, Mexico

ABSTRACT

Extensive cultivation of corn (*Zea mays* L.) for food, feed and biofuels may lead to a potential risk of soil degradation. Adoption of minimum tillage or no-till (NT) conservation systems help reduce this risk with benefits such as the increased residue cover, reduction of soil erosion and fuel consumption, and the improved efficiency of crop water use. All of these are important factors for sustainable crop production, especially in arid and semiarid regions. Although the production of corn under the NT system began approximately fifty years ago in the United States, and has become a more and more popular system worldwide since the 1990s, many aspects concerning the productivity improvement of conservation tillage have not been fully documented. One of the major concerns is the corn yield drag under cooler and wetter soil conditions in temperate humid regions due to the residue cover in no-till system, particularly in fine-textured clay loam or silt clay soils. It was often observed that reduction in early season nutrient uptake and early growth of the crop was associated with NT, resulting in the reduction of plant population density and grain yield. Use of starter fertilization and adoption of residue management in NT systems have been directed to alleviate the negative effects. Soil conditions of high moisture and increased rates of fertilization practices under NT are related to high N_2O emissions because they promote microbial denitrification; however it has been stated that agricultural N_2O emissions can be most effectively reduced with recommended practices to synchronize N application with crop demand. The purpose of this review

[*] Corresponding Author: B. L. Ma; E-mail: Baoluo.Ma@agr.gc.ca

is to summarize the results obtained in numerous studies carried out in different regions and to elucidate the main points of future research that allows a new approach to evaluate the impact of the fertilization management practices and the residue management measures on grain yield improvement and environment neutrality in a NT corn production system.

1. INTRODUCTION

Conservation tillage refers to any system that uses some tillage, but less than the conventional techniques of seedbed preparation. As one of the conservation tillage systems, no-till (NT) is defined as a system for planting crops without plowing and with use of herbicides to control weeds. The soil under NT is left undisturbed from harvesting of the previous crop to planting so as to minimize the disturbance and to maximize retention of crop residue on the surface. Seeding is usually done in one operation using an NT drill equipped with disk openers, hoe openers or coulters and with at least 30% mulch cover (Paulitz, 2006). Development of NT cropping systems, especially for corn production, began in the United States with the development of broad spectrum herbicides in the 1960s and was initially perceived as a technology to protect soil loss from wind and water erosion. It has since evolved into a system that offers some other benefits.

The production of corn (*Zea mays* L.) using NT management systems has continually increased in both temperate and tropical regions of the world. It is estimated that about 6% of cultivated crop lands in the world utilize NT systems, mostly in the United States, Canada, Australia, Argentina and Brazil (FAO, 2004). For example, in the United States, NT represents 19% of total cropped land (Triplett and Dick, 2008). Less wheel traffic in NT systems leads to lower soil compaction, thus improving soil structure and water infiltration, and promoting soil resilience in the tillage layer (Douglas and Goss, 1982; Dao, 1993; Kladivko, 2001; Lal et al., 2003). No-till system has also been promoted as a viable practice to increase soil carbon sequestration. However, the impact of NT system on corn grain yields and economic returns in different regions varies largely. De Felice (2006) stated that in the United States and Canada, yield response in NT is influenced by geography, soil conditions and crop rotation systems. The most important economic benefit in NT corn production is reduction in the use of gasoline and diesel fuels in farm machinery therefore reducing labor and capital requirements.

On the other hand, adoption of NT for corn production poses some problems compared to the conventional tillage system. Although crop residue cover under NT acts as a protection layer and thus reduces excessive runoff and soil erosion, leading to improved soil quality factors and increased efficiency of crop water use (Calegari, 2002; Erenstein 2003; Wildner, 2000; Grandy et al., 2006), it has been identified as the main factor for low temperatures and high moistures in the soil at planting time. As a consequence, grain yields of corn may be reduced due to soil physical limitations on root growth and distribution (Kaspar et al., 1987), and slow or uneven growth during the early stages and lower plant population densities (Dywer et al., 1995). Consequently, management of crop residues in the short and long term is one of the key factors for NT systems to succeed.

In addition to risk for negative soil impact, extensive cultivation of conventional tillage system is often thought to contribute to increased greenhouse gas (GHG) emissions.

Emissions of carbon dioxide (CO_2) and nitrous oxide (N_2O) from soil surface are the major components of the net global warming potential (Robertson and Grace, 2004). It was estimated that agriculture sector production contributes about 20% to the global annual increase in the atmospheric GHG concentration (IPCC, 2007). In this context, NT system has been promoted as one of the best management practices that increase soil organic carbon (C) storage or C sequestration. This process consists of transferring atmospheric CO_2 into long lived pools and keeping it stored in the soil securely so that it is not immediately re-emitted (Lal et al., 2003). Therefore, NT systems are regarded as a mitigation measure to reduce CO_2 emissions.

In contrast, conditions of high soil moisture in NT systems are related to high N_2O emissions because they promote microbial denitrification. It has been shown that there are higher populations of denitrifiers and higher denitrification rates for NT than for conventional tillage systems, and consequently there are often higher cumulative N_2O emissions reported, especially from poorly drained fine textured clay soils (Smith and Conen, 2004). It has been stated that agricultural N_2O emissions can be most effectively reduced with recommended practices to synchronize N application with crop demand. In the case of corn, some studies indicate that significant water and N uptake by a corn crop usually do not occur until around 8 weeks after seeding or up to one third of its life cycle (Ma and Dwyer, 1998; Ma et al., 2003). If high N rates are applied at planting, the potential for N losses could be considerable due to N_2O emissions in poorly drained soils and via NO_3^- leaching in well drained soils (Tan et al., 2009). Therefore, timing of N application is critical for minimizing N_2O losses, and it requires evaluating fertilization practices in NT corn production systems.

The placement of relatively small amounts of granular or soluble fertilizers in close proximity to the seed at planting which is called starter fertilization, could have an important role in mitigating agriculture contributions to global warming potential (GWP), either through CO_2 fixation by favouring rapid development of maximum leaf area during early vegetative stage, or through improved availability of nutrients to crop roots, thus minimizing NO_3^- leaching and curtailing N_2O emissions (Smith et al., 1997). Therefore, it is suggested that starter fertilization should be evaluated for its role in mitigating the negative environmental impacts of agriculture besides its effects on crop yields.

The objectives of this review are to present an overall picture of NT corn production, its pros and cons, to summarize the literature of fertilization technology as well as residue management measures that could overcome some of the limitations of NT systems on corn growth, and examine the potential of corn production in reducing the negative impact on the environment.

2. Effect of Crop Residue Cover in NT Corn Systems

One of the biggest benefits of NT crop system is the rapid development of a mulch cover on the soil surface. The vegetation mulch serves as the best protections against runoff and soil erosion due to wind and rainfall, and it reduces compaction caused for passing of heavy equipments or animals (Calegari, 2000; Erenstein, 2003; Seguy et al., 2003). In the long term, accumulation of crop residues due to continuous addition as well as their slow rates of decomposition enriches soil surface layer with decomposed organic matter and creates a favourable environment for beneficial soil fauna and flora activity. It is stated that increasing biological activity can also improve many important soil functions including improved fertilizer use efficiency, pesticide efficacy and C sequestration (Tripplet and Dick, 2008).

Residue cover is a factor in maintaining a high soil moisture as it helps reduce evaporation from the surface. Soil moisture conservation and retention is a benefit for no-till under dry conditions and on moderate- to well-drained soils as this can create a favourable environment for root development at/or near the mulch-soil interface (Triplett and Van Doren, 1969). Increased soil water content also helps attenuate extreme temperature fluctuations, which is beneficial for some soil functions related to the plant growth, especially in arid or semiarid regions, and in dry soils of tropical regions (Derpsch et al., 1986). However, in temperate regions high soil moisture due to surface residue cover has a negative effect as it will lead to low spring soil temperatures (Hayhoe et al., 1993). This effect is more severe on fine-textured, poorly drained soils than on coarse-textured sandy or loamy soils.

Quantitative assessment in corn production has shown that soil temperature is an important determinant of the growth from planting to the V_6 stage (Ritchie et al., 1993), when the apical growth point o f corn seedlings appears above the soil surface. Results of several studies lead to the conclusion of a linear relationship between shoot growth, root elongation or dry matter production and soil temperatures from 10 to 30 ^{o}C (Coelho and Dale, 1980). Other studies showed that root growth of corn ceases when the seedbed temperature is ≤ 9 ^{o}C (Warrington and Kanemasu, 1983; Swan et al., 1987). Hayhoe et al. (1993) reported that the effect of reduced soil temperatures during the emergence period was to increase the time required for corn seedling emergence, resulting in the reduced stand establishment due to longer seed exposure to soil pathogens. By comparing corn development under NT to conventional tillage systems, Fortin et al. (1994) suggest that accumulation of soil degree-days under NT could determine the total leaf area while optimal temperatures appear to be different for different leaves.

Lower soil temperature during the spring emergence period is associated with NT or reduced tillage compared with conventional tillage. Gupta et al. (1983) point out that the influence of surface residue cover is regulated by the energy available at the top soil layer, and by the reflective and thermal properties of the residue and the soil. Other studies confirm that large amounts of surface residues under NT conditions can act as an insulating layer to slow down the warming up of the soil after cold periods in temperate regions (Wall and Stobbe, 1983; Gupta et al., 1984; Hayhoe et al., 1993). Swan et al. (1987) demonstrated a linear relationship between in-row surface cover and number of growing-degree days required for seedling emergence, illustrating the quantitative impact of residue cover on the delay in plant growth.

The effect of low soil temperatures in the reduction of nutrient uptake in NT systems is regarded as one of the factors that lead to the slow growth of corn plants in the early stages. On one hand, low soil temperatures reduce the mineralization process, leading to the decreased availability of nutrients to young plant roots (Niehues et al., 2004); while on the other hand, the rate of root growth is also hampered by low soil temperatures. Mackay and Barber (1984) reported reduction of root growth as a result of poor P uptake in cool soils and accelerated root growth rates as soil temperatures rose from 18 to 25 ^{o}C. From a multi-site-year study, Dwyer et al. (2000) showed that soil temperatures at 12.5 ^{o}C or lower combined with high soil moisture (> 90% available water) reduced stand establishment to below 80%, which may be a factor in corn yield reductions associated with reduced or no tillage. Low soil temperatures during early growth may also affect the metabolism and morphology of roots (Stamp et al., 1997).

Besides the effects of NT on high soil moisture and low temperature in the early growth stages, root morphology can also be modified due to the greater soil strength and bulk density in NT systems. In general, NT increases soil bulk density, lowers saturated hydraulic conductivity, reduces air filled porosity and increases soil strength when compared with conventional tillage systems (Pierce et al., 1992). Chassot at al. (2001) indicate that restriction of root growth under NT due to soil compaction is limited, and delayed root growth is mostly caused by lower temperatures than soil physical restrictions; and such an effect is more pronounced in fine-textured soils. Dwyer et al. (1995) found that physical conditions of soil under NT did not affect the root mass distribution, but influenced the rooting depth.

Depending upon soil type and environmental conditions, changes in vertical and horizontal distribution of soil fertility can occur over time in continuous NT systems. Duiker and Beegle (2006) observed strong pH vertical gradient in a site under long term NT, but such changes in soil pH gradient did not occur in another site with the same soil type (Luvisol) and similar NT history but with more rainfall (350 mm yr^{-1} more). Mineralization of organic matter from crop residues, application of ammonium fertilization, and root exudation are some of the acidification factors in the top soil layer in NT (Limousin and Tessier, 2007).

Phosphorous and K nutrients in soils under NT conditions tended to concentrate in the layer of 0 to 5 cm depth after several years of fertilization. The top soil generally has higher CEC values under NT, which could explain the high Ca levels in the same layer when the soil was not limed prior to the study (Duiker and Beegle, 2006). According to Kapusta et al. (1996), the impact of residue cover on the redistribution of nutrients has not proven detrimental to root growth or corn yield since it has been found that root extraction of water and nutrients is greater with NT than conventional tillage.

3. Crop Residue Management in NT Systems

Despite the benefit of crop residues on the protection against soil erosion, there are also negative effects associated with low soil temperatures, high soil moisture and increased soil strength on the growth of roots. Consequently, crop production in NT systems often suffers from other limitations such as erratic germination, poor stand establishment, poor early vigour, and difficulty in weed control (Swan at al., 1994). The degree of impaction varies according to the type of previous crop, quantity and time of crop residue accumulation, as well as soil and climatic conditions. Therefore, appropriate residue management practices in NT corn systems are required to alleviate these effects.

There are two major schemes for the residue management in NT systems: crop residues are unremoved during planting or residues are partially removed from the seed row. Common names for the latter are strip tillage, zone tillage or in-row residue management.

Benefits of residue removal from seed row in NT systems have been reported (Kaspar et al., 1990). Removal of crop residues from the seed row can reduce the amount of time needed to emergence and tasseling by 50%, as a result, compared with conventional tillage, NT treatment increased plant height, decreased grain moisture, and minimized the yield-drag associated with NT system when a 16 cm band centered on the seed row was kept residue free (Kaspar et al., 1990).

Despite the positive effect of crop residue removal on emergence and early growth, most studies have not found significant or consistent impact of in-row residue removal on grain yields. For example, Swan et al. (1994) concluded that in-row residue removal has no effect on yield despite long-term (seven years) changes in soil physical and chemical characteristics. Similarly, by comparing in-row residue removal against without residue removal both under NT system, Fortin (1993) showed the beneficial effect of in-row crop residue removal on the increased leaf number, plant height and soil temperatures, but not on soil water content. He did not observe any differences in grain yield between the two practices. Azooz et al. (1995) reported that residue removal in a band of 30 cm over the seed row at planting resulted in soil thermal and moisture conditions that were equivalent to a conventional tillage system, and residue removal improved emergence and early growth rates but had no effect on corn yield. Wolkowski (2000) observed that the soil was drier and warmer for in-row residue removal than NT system without residue removal. In that case, the level of crop residues was about 30-50% in the row and 80-90% in the inter-row of NT without residue removal, which offered an important surface protection against erosion and improved emergence and early growth. However, there was again no grain yield improvement. Janovicek et al. (1997) found that clearing in-row residues increased early season corn growth for rotational corn, but not for continuous corn monoculture, both with NT system. Therefore, in-row removal of crop residues cannot always be expected to overcome the yield limiting effects associated with continuous NT corn production.

4. Nutrient Management in NT Corn Production

In general, N fertilizer requirement in NT is greater than in conventional tillage systems. There is low incorporation of fertilizers into the top layer in NT systems, resulting in the limited contact or mixing between the soil and the surface-applied fertilizers. Nitrogen immobilization under NT is also increased due to the slower process of crop residue decomposition. Soil physical conditions under NT may limit movement of nutrients towards the plant root system by diffusion (P, K) or by mass flow (NO_3^--N, Ca, S); uptake of these nutrients by the root from the soil solution is therefore limited (Mengel, 1995; Torbert et al., 2001). As a result, shift from conventional tillage towards NT systems for corn production may require higher rates of N fertilization to compensate the limitations. On the long run, NT production may alter nutrient distribution as well as some chemical and physical conditions, making it necessary to implement best fertilization practices including placement, amount and source of nutrients for different soils (Franzluebbers and Hons, 1996).

The main purpose of starter fertilization is to improve the efficiency by targeting fertilizers more accurately in relation to crop roots. In general, starter fertilizer placement methods refer to applying and incorporating the fertilizer into a large zone of soil (broadcasting) or, in a very restricted band of soil. The latter increases nutrient availability to the crop roots by increasing the concentration of nutrients in a small volume of soil, which is important, especially for nutrients relatively immobile such as P and K (Randal and Hoeft, 1988). Use of starter fertilization has been suggested as a viable strategy with greater precision in NT corn production systems as this practice provides the crop roots with readily available nutrients at early growth stages. Starter fertilization is often recommended as a strategy to reduce overall fertilizer use, thereby curtailing the negative impact on environment, for several crops in both conventional tillage and NT production systems (Stone at al., 1999).

Starter fertilizers can be granular or liquid mixtures that usually are two or more combinations of N-P-K at different ratios. Starter fertilizers sometimes also contain other nutrients such as S or Zn. Responses of corn to starter fertilization have become a topic of interest for the purpose of enhancing corn productivity or preventing yield drag in NT systems. It has been widely shown that corn performance with starter fertilization in NT systems depends on soil moisture, temperature (Mackay and Barber, 1984; Fortin, 1993; Niehues et al., 2004) and climatic factors. Some studies suggest that N is the most important nutrient for producing a yield response, while others indicate that P is also important. While the quantitative response of corn crop to each nutrient starter fertilizer has not been well established (Scharf, 1999; Gordon and Pierzynski, 2006; Wortmann et al, 2006), it is generally observed that starter fertilization improves early growth and stand establishment of corn crop.

The method of fertilizer placement is an important aspect of starter fertilization strategy in NT corn system as the contact between plants and fertilizers is often hampered by the residue cover. There are two common methods of placement in NT systems: place the fertilizer (i) at near the seeds in shallow bands (usually at 5 cm to the side and 5 cm below the seed row) or (ii) with the seed. Placement with the seed (also referred as to in-furrow or pop-up) requires fertilizers with low salts to avoid risk of salt injury; urea and diammonium phosphate are not recommended as these fertilizers release ammonia, which can be toxic to the germinating seeds (Bermudez and Mallarino, 2002). The effect of fertilizer placement methods on crop yields is often shown to be influenced by soil texture, soil moisture and source of fertilizers, while depth of fertilizer placement is often reported not to have a consistent effect. For example, Reeves et al. (1986) reported no difference between shallow and deep placements of N and P in both silt and sandy textured soils. Vyn et al. (2000) found that there were no significant corn yield benefits with deep banding of K fertilizer relative to shallow placement. In soils with high or low test P or K, Bardoli and Mallarino (1988) observed that deep band placement improved P uptake and yield but their results for K were inconsistent. Kaspar et al. (1990) suggest that subsurface placement may be preferable to surface broadcast application to prevent susceptibility of crop roots to soil drying. Mengel et al. (1988) and Wortman et al. (2006) reported similar effects of in-furrow and band placement on crop performance.

In conservation tillage systems, surface application of P and K fertilizers and annual return of corn residues can result in the stratification of these nutrients in the topsoil (Mackay et al., 1987). In order to provide growing roots with sufficient P and K, it may be necessary to apply fertilizers with deep placement. However, a positive response of corn yield to P and K fertilization could not be expected in soils with relatively high soil test values (Mengel et al., 1988). Bardoli and Mallarino (1998) point out that potential problem for proper management of P and K in NT corn systems include inappropriate extrapolation of soil test interpretations from conventional tillage, inappropriate soil sampling techniques and inefficient fertilizer placement. For soils with high P levels, Roth et al. (2006) suggest the use of high N rates or in-furrow application of starter fertilizer. Similarly, for soils with high test K, Riedell et al. (2000) showed the importance of application of P fertilizer at preplant for optimum growth and yield of corn hybrids, while grain yields will not be improved with additional K fertilizer or deep placement of K in irrigated NT systems.

Various N-P-K combinations of starter fertilization have been used for NT corn production. Responses of corn growth and yield to combinations of starter fertilization are dependent on environmental conditions, including soil texture, manure history, preceding crop and soil moisture. Gordon and Pierzynski (2006) reported that in a sandy loam soil, combinations of N-P or N-P-K-Zn consistently increased grain yields, reduced the number of thermal units needed from emergence to maturity, decreased grain moisture content at harvest and increased total P uptake in one of the two tested hybrids, though K and Zn did not show any additional benefit. In general, inclusion of N in starter fertilizer will have a positive effect on N and P uptake, plant height and early vigour during the early growth stages; but these early responses may not always be translated into greater grain yields. Some studies reported increases in grain yield as a result of application of different combinations of starter fertilization (Revees at al., 1986; Touchton and Karim, 1986; Mengel, 1990; Ritchie, 1996; Shulte and Bundy, 1994; Scharf, 1999) while others observed no grain yield response (Kapusta and Varsa, 1990; Kapusta et al., 1996).

Dick et al. (1992) found that the addition of a small amount of N (56 or 112 kg ha^{-1}) stimulated organic matter decomposition, but only split applications of high fertilizer rates (336 kg N ha^{-1}) increased grain yields. In the latter case, crop N uptake under NT system was significantly greater than that of conventional tillage. In a sandy loam soil with high levels of soil test P and K, Touchton and Karim (1986) observed that differences in plant height were closely related with differences in grain yields, which was associated with soil N supply levels, but not with differences in nutrient concentrations in leaf tissues.

Different responses of hybrids to starter N fertilization have been attributed to differences in root-growth characteristics of the hybrids (Gordon and Pierzynski, 2006), amounts of residue cover and sources of N supply. In this regard, Viswakumar et al. (2008) indicate that surface broadcast applications of urea ammonium nitrate in NT corn production can lead to high rates of ammonia volatilization and low grain yields. Similarly, Andraski and Bundy (2008) observed that regardless of N sources, increasing the amount of crop residue led to lower soil temperature and lower soil NO_3^--N concentrations in the early season. Grain yield was also affected by N sources (ammonium nitrate, urea ammonium nitrate, and urea treatments) in their study.

5. Environmental Impact of NT Systems

Atmospheric N_2O concentration has increased from about 273 ppb in 1750 to 319 ppb in 2005, a concentration increase of 0.2 ppb yr^{-1}, and more recently, at the unprecedented increase of 0.8 ppb yr^{-1} (IPCC, 2007). Smith and Conen (2004) reported that 20% of the annual global increase in GHG concentrations arose from agriculture practices, 25% of which were N_2O emissions from agricultural lands. Soil emitted N_2O is the by-product of both the nitrification and denitrification processes, which are mediated by obligate chemoautotrophic bacteria and facultative anaerobes, respectively (Davidson and Swank, 1986; de Boer et al., 1988). Biological denitrification, the main source of N_2O emissions refers to the reduction of NO_3^- to N_2 by soil bacteria in the absence of oxygen and where N_2O is generated as an intermediate volatile compound (Duxbury et al., 1982; Tan et al., 2009).

The rate and total amount of N_2O production is controlled by soil physical and chemical properties (Beauchamp, 1997; Bouwman and Boumans, 2002; Davidson and Swank, 1986; Smith et al., 2003), including soil moisture and soil aeration, NH_4^+ and NO_3^- concentrations, and soil C levels that affect the growth of denitrifier populations and then the flux of N_2O emission (Bacon and Freney, 1989; Dobbie and Smith, 2003; Fortuna et al., 2003; Smith et al., 1998a,b; Tan et al., 2009). Nitrogen fertilization is one of the main sources of anthropogenic N_2O emission to the atmosphere (Cole et al., 1997). The effect of N fertilization on N_2O emission from the soil has been described as a hole in a pipe: the more N fertilizer is applied in the soil, the greater the amount converted to N_2O and released to the atmosphere (Smith et al., 1997). Therefore, recent studies are focusing on assessing the role of different N fertilization strategies in the intensity of N_2O emissions, such as N rates, timing, and placement as well as residue managements in NT corn production systems.

Bowman (1996) suggested a linear relationship between the rate of N fertilizer applied and N_2O emissions. It is defined by a regression equation where it is also considered a background flux caused by the cycling of N from crop residues, soil organic matter, atmospheric deposition, and previous fertilization in cultivated tilled soils. This relationship clearly indicates that the more fertilizer applied, the greater the N_2O emission. In general, the IPCC method of estimating the contributions of fertilizers to global anthropogenic N_2O fluxes (IPCC, 2001) assumed fertilizer-induced emission at a range of $1.25\pm1\%$ of the N applied. However, more recent studies suggest that fertilizer-induced N_2O emissions from agricultural fields range from $< 0.5\%$ to $> 2\%$ of the N fertilizer applied (Stehfest and Bowman, 2006; Adviento-Borbe et al., 2007; Ma et al., 2009).

Site specific factors can greatly affect the value of fertilizer-induced emissions. Sehy et al. (2003) found that an increase of N input by 17% in high corn yielding sites affected neither crop yield nor N_2O emissions. In contrast, reducing N input by 17% in the low yielding area resulted in 34% less N_2O emission losses without affecting the yield. With soil N below the threshold for elevated N_2O emissions (3.5 mg NO_3^--N kg^{-1} d.wt.), high soil water content rather than N fertilizer rates was responsible for increasing N_2O emissions. In an on-farm study on the effect of the amount and timing of N fertilizer application to corn, Ma et al. (2009) recently reported that the cumulative N_2O emissions ranged from 0.05 to 2.42 kg N ha^{-1}, equivalent to 0.03 to 1.45% of the applied N during the initial 28-d monitoring period after preplant plus a second 28-d monitoring period after sidedress application. On average, the cumulative mean N_2O emission averaged 0.26 kg N ha^{-1} for the unfertilized control treatment, but increased more than two fold to 0.78 kg N ha^{-1} with an additional 60 kg N ha^{-1} sidedress on the base of 30 kg N ha^{-1} starter fertilizer. By increasing the sidedress N to 120 kg N ha^{-1} the cumulative N_2O emissions increased by twofold again to 1.30 kg N ha^{-1}. Their study clearly suggests that any increase in N input beyond 90 kg N ha^{-1} would result in considerable N_2O emissions with marginal yield improvement in the cool and humid regions.

It has been recognized that proper nutrient management practices will reduce N_2O emissions substantially. For example, in high-yielding corn production systems, Adviento-Borbe et al. (2007) showed that N_2O emissions were not directly related to the N rates, even with increases of N fertilizer rates by 40% in continuous corn or by 64-92% in corn following soybean than the recommended corresponding rates. They concluded that the majority of the N_2O was released during the growing season and that soil temperatures and cumulative precipitation prior to sampling were the main factors correlated with N_2O emission rates.

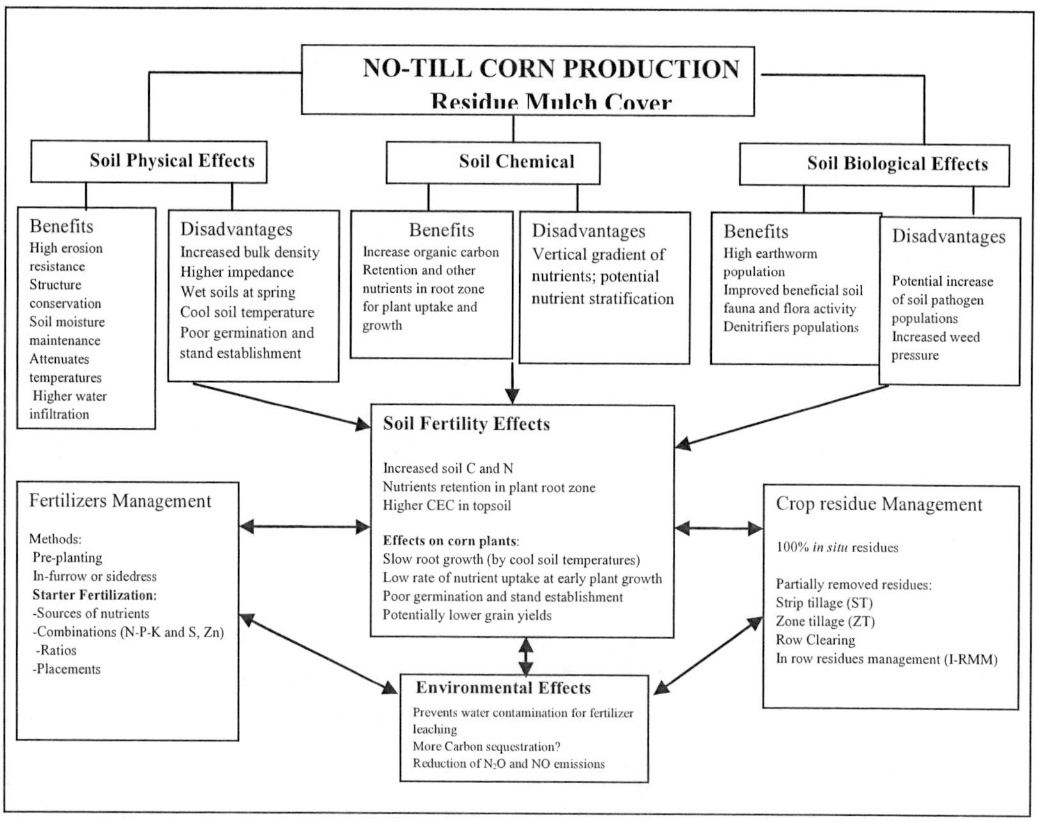

Figure 1. A diagram showing the impact of crop residue and fertilizer managements on soil fertility and the environment in a NT corn production system.

While both nitrification and denitrification can produce soil N_2O emissions, it is denitrification that results in the greatest amount of N_2O emissions, especially when soil aeration is poor (Chantigny et al., 1998). The effect of soil moisture on the N_2O emissions has also been shown in NT crop production systems. Sehy et al. (2003) reported a value of 60% water filled pore space (WFPS) as a threshold for the induction of elevated N_2O emissions. They observed a significant relationship between N_2O flux and soil WFPS when soil mineral N was not limiting. Similarly, Adviento et al. (2007) reported that N_2O emission peaks were mainly associated with high temperatures and high soil water content resulting from rainfall or irrigation events rather than soil NO_3^--N levels in high yielding corn systems. Tan et al. (2009) observed that N_2O emissions peaked on the second day after a simulated rainfall event regardless of N fertilizer treatments; N_2O losses were significantly higher on the clay loam compared to the sand loamy soils. Fertilizer placement practices particularly due to longer exposure of fertilizer in surface of land compared with conventional tillage also directly affect the N_2O flux rates (Smith et al., 1997; Viswakumar et al., 2008).

Crop residues in NT systems are a source of organic C for soil microorganisms and contribute to plant nutrients. Figure 1 is a diagram showing how the crop residue cover under NT systems affects soil physical, chemical and biological processes and the impact of crop residue and fertilization management. Corn residues like any other cereal crop residues contribute little immediate plant available N to the soil because of their high C:N ratios (Malhi et al., 2006). With high soil moisture and anaerobic conditions, C substrate from the residue favours denitrification microbial activity, producing N_2O and NO that are released to the atmosphere.

In contrast, conventional tillage with mouldboard plow loosens the soil and allows temperature to rise quickly in the spring. As a result, soil organic matter is decomposed through mineralization/ nitrification microbial activities, resulting in direct release of CO_2 into the atmosphere and increased soil mineral N. Breakdown of soil organic matter may also lead to accelerated losses of N_2O and NH_3 emissions and NO_3^- leaching under cool and saturated soil conditions (Malhi et al., 2006).

Timing of N application is critical for minimizing N losses under NT systems. Tan at al. (2009) reported that early fertilizer application at full rates were susceptible to losses through leaching and denitrification following a rainfall event (50 mm), while there was much lower net losses if fertilizer is applied as the starter. However, the differences are strongly related with soil texture as they found that denitrification-related N_2O emissions were four times higher for clay loam soil than for loamy sand soil. Therefore, adoption of nutrient best management practices such as adjusting the rates of N applied to corn in NT systems and matching the timing of N application as close as possible to crop demand, is also likely to be beneficial in reducing N_2O emissions.

6. Summary and Concluding Remarks

Extensive production of corn under NT system is an option to diminish the risk of soil erosion, however, adoption of NT poses some problems related to plant growth, grain yield, crop residues, and physical and chemical soil conditions that favoured negative environmental processes. Most of the studies on the management of fertilization and crop residues in NT corn production have focused on assessing the effects of NT on the crop growth and grain yields. Alternative measures are therefore proposed to improve specific factors influencing corn plant development in specific regions. Since corn cultivation is extended on a wide range under different climates and soil types, the response to proposed treatments could be beneficial in some areas but not in others. Therefore, reduction of early nutrient uptake and early growth associated with NT and the resulting reduction of plant population density and grain yield are still some important aspects that continue to be investigated for specific regions.

Currently, increasing GHG concentrations in the atmosphere has raised considerable concerns about emissions of CO_2 and N_2O from soil surface of agricultural lands. It is stated that NT system offers some environmental benefits compared with conventional tillage systems, such as higher rates of C sequestration in the top soil layer where mulch residues are formed. However, high moisture conditions and fertilization practices may promote denitrification processes that could result in high rates of N_2O emissions to the atmosphere. Therefore, it is of crucial importance to implement the more convenient methods of placement, timing and, fertilizer combinations for specific sites in order to improve their efficiency and diminish the environmental impact.

Given the effects of crop residue cover on the soil conditions, fertilizer practices such as starter fertilization has the potential to improve the corn productivity in NT systems when associated with appropriate crop residue management. It is suggested that in the future both starter fertilization methods and crop residue managements should be evaluated as associated components of NT system in order to obtain maximum yield potential and long term sustainability.

Yield production and environmental impact evaluations from NT corn production systems should standardize the criteria for measuring parameters such as N-fertilizer uptake, C removal in crops, depth of organic C accumulation, and fluxes of N_2O and CO_2 in the system in order to compare with other tillage systems.

References

Adviento-Borbe, M.A.A., Haddix, M.L., Binder, D.L., Walters, D.T., Dobermann, A. 2007. Soil greenhouse gas fluxes and global warming potential in four high-yielding maize systems. Global Change Biol. 13: 1972-1988.

Andraski, T.W. and Bundy, L.G. 2008. Corn residue and nitrogen sources effects on nitrogen availability in no-till corn. Agron. J. 100: 1274-1279.

Azooz, R.H., Lowery, B. and Daniel, T.C. 1995. Tillage and residue management influence on corn growth. Soil Till. Res. 33: 215-227.

Bacon, P.E. and Freney, J.R. 1989. Nitrogen loss from different tillage systems and the effect on cereal grain yield. Fertilizer Res. 20: 59-66.

Bardoli, J.M. and Mallarino, A.P. 1998. Deep and shallow banding of phosphorus and potassium as alternatives to broadcast fertilization for no-till corn. Agron. J. 90: 27-33.

Beauchamp, E.G. 1997. Nitrous oxide emission from agricultural soils. Can. J. Soil Sci. 77: 113–123.

Bermudez, M. and Mallarino, A.P. 2002. Yield and early growth responses to starter fertilizer in no-till corn assessed with precision agriculture technologies. Agron. J. 94: 1024-1033.

Bouwman, A.F. 1996. Direct emission of nitrous oxide from agricultural soils. Nutr. Cycl. Agroecosyst. 46: 53-70.

Bouwman, A.F. and Boumans, L.J.M. 2002. Emissions of N_2O and NO from fertilized fields: Summary of available measurement data. Global Biogeochem. Cycl. 16: 1-13.

Calegari, A. 2002. The spread and benefits of no-till agriculture in Parana State Brazil. *In* Agroecological Innovations: Increasing Food Production with Participatory Development. N. Uphoff, Ed.: 187-202. Earthscan, London.

Chantigny, M.H., Prévost, D., Angers, D.A., Simard, R.R. and Chalifour, F.P. 1998. Nitrous oxide production in soils cropped to corn with varying N fertilization. Can. J. Soil Sci. 78: 589–596.

Chassot, A., Stamp, P. and Richner, W. 2001. Root distribution and morphology of maize seedlings as affected by tillage and fertilizer placement. Plant Soil 231: 123-135

Coelho, D.T. and Dale, R.F. 1980. An energy-crop growth variable and temperature function for predicting corn growth and development: Planting and silking. Agron. J. 72: 503-510.

Dao, T.H. 1993. Tillage and winter wheat residue management effects on water infiltration and storage. Soil Sc. Soc. Am. J. 57: 1586-1595.

Davidson, E.A. and Swank, W.T. 1986. Environmental parameters regulating gaseous nitrogen losses from two forested ecosystems via nitrification and denitrification. Appl. Environ. Microbiol. 52: 1287-1292.

de Boer, W., Duyts, H. and Laanbroek, H.J. 1988. Autotrophic nitrification in a fertilized acid heath soil. Soil Biol. Biochem. 20: 845-850.

de Felice, M. S., Carter, P. R. and Mitchell, S. B. 2006. Influence of tillage on corn and soybean yield in the United States and Canada. Website available at www.plantmanagementnetwork.org/cm

Derpsch, R. 1986. Erosion problems in Parana, Brazil. Research results and strategies for the implementation of efficient soil conservation measures. M.Sc. Thesis. Agricultural Extension and Rural Development Centre, University of Reading. UK.

Dick, W.A., Edwards, W.M., Stehouwer, R.C. and Eckert, D.J. 1992. Maize yield and nitrogen uptake after established no-tillage fields are plowed. Soil Till. Res. 24:1-10.

Dobbie, K.E. and Smith, K.A. 2003. Nitrous oxide emission factors for agricultural soils in Great Britain: the impact of soil water-filled pore space and other controlling variables. Global Change Biol. 9: 204-218.

Douglas, J.T. and Goss, M.J. 1982. Stability and organic matter content of surface soil aggregates under different methods of cultivation and in grassland. Soil Till. Res. 2: 155-175.

Duiker, S.W. and Beegle, D.B. 2006. Soil fertility distributions in long-term no-till, chisel/disk and mouldboard plow/disk systems. Soil Till. Res. 88: 30-41.

Dwyer, L.M., Ma, B.L., Hayhoe, H.N. and Culley, J.L.B. 1995. Tillage effects on soil temperature, shoot dry matter accumulation and corn grain yield. J. Sustainable Agric. 5: 85-99.

Dwyer, L.M., Ma, B.L., De Jong, R. and Tollenaar, M. 2000. Assessing corn seedbed conditions for emergence. Can. J. Soil Sci. 80: 53-61.

Erenstein, O. 2003. Smallholder conservation farming in the tropical and sub-tropic: A guide to the development and dissemination of mulching with crop residues and cover crops. Agric. Ecosyst. Environ. 100: 17-37.

FAO. 2004. FAOSTAT. FAO, Rome. Available online at //www.fao.org/corp.statistics/

Fortin, M.C. 1993. Soil temperature, soil water, and no-till corn development following in-row residue removal. Agron. J. 85: 571-576.

Fortin, M.C., Pierce, F.J. and Edwards, M. 1994. Corn leaf area response to early-season soil temperature under crop residues. Agron. J. 86: 355-359.

Fortuna, A., Harwood, R.R., Robertson, G.P., Fisk, J.W. and Paul, E.A. 2003. Seasonal changes in nitrification potential associated with application of N fertilizer and compost in maize systems of southwest Michigan. Agric. Ecosyst. Environ. 97: 285-293.

Franzluebbers, A.J. and Hons F.M. 1996. Soil profile distribution of primary and secondary plant-available nutrients under conventional and no tillage. Soil Till. Res. 39: 229-239.

Gordon, W.B. and Pierzynski, G.M. 2006. Corn hybrid response to starter fertilizer combinations. J. Plant Nutr. 29: 1287-1299.

Grandy, A.S., Robertson, G.P. and Thelen, K.D. 2006. Do productivity and environmental trade-offs justify periodically cultivating no-till cropping systems? Agron. J. 98: 1377-1383

Gupta, S.C., Larson, W.E. and Linden, D.R. 1983. Tillage and surface residue effects on soil upper boundary temperatures. Soil Sci. Soc. Am. J. 47: 1212-1218.

Gupta, S.C., Larso,W.E. and Allmaras, R.R. 1984. Predicting soil temperature and soil heat flux under different tillage-surface residue conditions. Soil Sc. Soc. Am. J. 48: 223-232.

Hayhoe, H.N., Dwyer, L.M., Balchin, D. and Culley, J.L.B. 1993. Tillage effects on corn emergence rates. Soil Till. Res. 26: 45-53.

IPCC. 2007. Climate Change and Water: The Physical Science Basis. Report of Intergovernmental Panel on Climate Change 2007. Available online at http://www.ipcc.ch/

IPCC. 2001. Climate Change 2001. The Scientific Basis. Report of Intergovernmental Panel on Climate Change 2001. Cambridge University Press, Cambridge, UK.

Janovicek, K.J., Vyn, T.J. and Voroney, R.P. 1997. No-till corn response to crop rotation and in row residue placement. Agron. J. 89: 588-596.

Kladivko, E. 2001. Tillage system and soil ecology. Soil Till. Res. 61: 61-76.

Kaspar, T.C., Crosbie, T.M., Cruse, R.M., Erbach, D.C., Timmons, D.R. and Potter, K.N. 1987. Growth and productivity of four hybrids as affected by tillage. Agron. J. 79: 477-481.

Kaspar, T.C., Erbach, D.C. and Cruse, R.M. 1990. Corn response to seed-row residue removal. Soil Sci. Soc. Am. J. 54: 1112-1117.

Kapusta, G. and Varsa, E.C. 1990. Long term tillage and fertility effects in corn. *In* Proceedings reports of the North Central Extension – Industry Soil Fertility Conference. p70-81. Potash and Phosphate Institute (ed.). Manhattan, KS.

Kapusta, G., Krausz, R.F. and Matthews, J.L. 1996. Corn yield is equal in conventional, reduced and no tillage after 20 years. Agron. J. 88: 812-817.

Lal, R., Follett, F. and Kimble, J.M. 2003. Achieving soil carbon sequestration in the United States: A challenge to the policy makers. Soil Sci. 168: 827- 845.

Limousin, G. and Tessier, D. 2007. Effects of no-tillage on chemical gradients and topsoil acidification. Soil Till. Res. 92: 167-174.

Ma, B.L. and Dwyer, L.M. 1998. Nitrogen uptake and utilization of two contrasting maize hybrids differing in leaf senescence. Plant Soil 199: 283-291.

Ma, B.L., Ying, J., Dwyer, L.M., Gregorich, E.G. and Morrison, M.J. 2003. Crop rotation and soil nitrogen amendment effects on maize production in Eastern Canada. Can. J. Soil Sci. 83: 483-495.

Ma, B.L., Wu, T.Y., Tremblay, N., Deen, W., Morrison, M.J., McLaughlin, N.B., Gregorich, E.G. and Stewart, G. 2009. Nitrous oxide fluxes from corn fields: on-farm assessment of the amount and timing of nitrogen fertilizer. Global Chang Biol. (*In press*). doi: 10.1111/j.1365-2486.2009.01932.x.

Mackay, A.D. and Barber, S.A. 1984. Soil temperature effects on root growth and phosphorus uptake by corn. Soil Sc. Soc. Am. J. 48: 818-823.

Mackay, A.D., Kladivko, S.A., Barber, S.A. and Griffith, R.D. 1987. Phosphorus and potassium uptake by corn in conservation tillage systems. Soil Sci. Soc. Am. J. 51: 970-974.

Malhi, S.S., Lemke, R., Wang, Z.H. and Chhabra, B.S. 2006. Tillage, nitrogen and crop residue effects on crop yield, nutrient uptake, soil quality, and greenhouse gas emissions. Soil Till. Res. 90:171-183.

Mengel, D.B., Hawkins, S.E. and Walker, P. 1988. Phosphorus and potassium placement for no-till and spring plowed corn. J. Fert. Issues 5 (1): 31-36.

Mengel, D.B. 1995. Roots, growth and nutrient uptake. Dept. Agron. Publ. AGRY 95-08. 8pp. Purdue University Cooperative Extension Service, West Lafayette, IN.

Mengel, D.B. 1992. Fertilizing corn grown using conservation tillage. Agron. guide 268. Purdue University Cooperative Extension Service, West Lafayette, IN.

Niehues, B.J., Lamond, R.E., Godsey, C.B. and Olsen, C.J. 2004. Starter nitrogen fertilizer management for continuous no-till corn production. Agron. J. 96: 1412-1418.

Paulitz, T.C. 2006. Low input no-till cereal production in the Pacific Northwest of the U.S.: The challenge of root diseases. Eur. J. Plant Pathol. 115: 271-281.

Pierce, F.J., Fortin, M.C. and Staton, M.J. 1992. Immediate and residual effects of zone-tillage in rotation with no-tillage on soil physical properties and corn performance. Soil Till. Res. 24: 149-165.

Randall G.W. and Hoeft, R.G. 1988. Placement methods for improved efficiency of P and K fertilizers: A review. J. Prod. Agric. 1: 70-79.

Reeves, D.W., Touchton, J.T. and Burmester, C.H. 1986. Starter fertilizer combinations and placement for conventional and no-tillage corn. J. Fert. Issues 3 (3): 80-85.

Riedell, W.E., Beck, D.L. and Schumacher, T.E. 2000. Corn response to fertilizer placement treatments in an irrigated no-till system. Agron. J. 92: 316-320.

Ritchie S.W., Hanway, J.J. and Benon, G.O. 1993. How a maize plant develops. Sp. Rpt. No. 48. Iowa State University of Science and Technology. Co-operative Extension Services. Ames. IA.

Robertson, G.P. and Grace, P.R. 2004. Greenhouse fluxes in tropical and temperate agriculture: the need for a full-cost accounting of global warming potentials. Environ. Develop. Sustain. 6:51-63.

Roth, G.W., Beegle, D.B., Heinbaugh, S.M. and Antle, M.E. 2006. Starter fertilizers for corn on soils testing high in phosphorus in the Northeastern USA. Agron. J. 98: 1121-1127.

Scharf, P.C. 1999. On-farm starter fertilizer response in no-till corn. J. Prod. Agric. 12: 692-695.

Seguy, L., Bouzinac, S., Scopel, E. and Ribeiro, M.F.S. 2003. New concepts for sustainable management of cultivated soils through direct seeding mulch based cropping systems: The CIRAD experience, partnerships and networks. 142-144. *In* Proceedings of the II World Congress on Conservation Agriculture-Producing in Harmony with Nature. August 2003, Foz do Iguaçu, Brazil.

Sehy, U., Ruser, R. and Munch, J.C. 2003. Nitrous oxide fluxes from maize fields: relationship to yield, site-specific fertilization, and soil conditions. Agric. Ecosys. Environ. 99: 97-111.

Smith, K.A., Ball, T., Conen, F., Dobbie, K.E., Massheder, J. and Rey, A. 2003. Exchange of greenhouse gases between soil and atmosphere: interactions of soil physical factors and biological processes. Eur. J. Soil Sci. 54: 779-791.

Smith, K.A. and Conen, F. 2004. Impacts of land management on fluxes of trace greenhouse gases. Soil Use Manage. 20: 255-263.

Smith, K.A., McTaggart, I.P. and Tsuruta, H. 1997. Emissions of N_2O and NO associated with nitrogen fertilization in intensive agriculture, and the potential for mitigation. Soil Use Manage. 13: 296-304.

Smith, K.A., McTaggart, I.P., Dobbie, K.E. and Conen, F. 1998a. Emissions of N_2O from Scottish agricultural soils, as a function of fertilizer N. Nutr. Cycl. Agroecosyst. 52: 123-130.

Smith, K.A., Thomson, P.E., Clayton, H., McTaggart, I.P. and Conen, F. 1998b. Effect of temperature, water content and nitrogen fertilization on emissions of nitrous oxide by soils. Atmospheric Environ. 32: 3301-3309.

Stamp, P., Feil, M., Shortemeyer, B. and Richner, W. 1997. Responses of roots to low temperatures and nitrogen forms. p.143-154. *In* H.M. Anderson et al. (ed.) Plant roots to cells to systems. Kluwer Academic Publ., Dordrecht, The Netherlands.

Stone, D.A., Tyler, F.T. and Runham, S. 1999. Effect of starter fertilizers on the yield and quality of lettuce grown on peaty soils. Soil Use Manage. 15: 128-132.

Swan, J.B., Schneider, E.C., Moncrief, J.F., Paulson, W.H. and Peterson, A.E. 1987. Estimating corn growth, yield, and grain moisture from air growing degree days and residue cover. Agron. J. 79: 53-60.

Swan, J.B., Higgs, R.L., Bailey, T.B., Wollenhaupt, N.C., Paulson, W.H. and Peterson, A. 1994. Surface residue and in-row treatment effects on long term no-tillage continuous corn. Agron. J. 86: 711-718.

Tan, I.Y.S., van Es, H.M., Duxbury, J.M., Mekonian, J.J., Schindelbeck, R.R., Geohring, L.D., Hively, W.D. and Moebius, B.N. 2009. Single-event nitrous oxide losses under maize production as affected by soil type, tillage, rotation and fertilization. Soil Till. Res. 102: 19-26.

Torbert, H.A., Potter, K.N. and Morrison, J.E., Jr. 2001. Tillage system, fertilizer nitrogen rate, and timing effect on corn yields in the Texas Blackland prairie. Agron. J. 93: 119-1124.

Touchton, J.T. and Karim, F. 1986. Corn growth and yield responses to starter fertilizers in conservation tillage systems. Soil Till. Res. 7: 135-144.

Triplett, G.B. and Van Doren, D.M. 1969. Nitrogen, phosphorus, and potassium fertilization of non- tilled maize. Agron. J. 61: 637-639.

Triplett, G.B. and Dick, W.A. 2008. No-tillage crop production: A revolu tion in agriculture! Agron. J. 100: S153-S165.

Viswakumar, A., Mullen, R.W., Sundermeier, C.E. and Dygert, C.E. 2008. Tillage and nitrogen application methodology impacts on corn grain yield. J. Plant Nutr. 31: 1963-1974.

Vyn, T.J., Galic, D.M. and Janovicek, K.J. 2002. Corn response to potassium placement in conservation tillage. Soil Till. Res. 67: 159-169.

Wall, D.A. and Stobbe, E.H. 1983. The response of eight corn (*Zea mays* L.) hybrids to zero tillage in Manitoba. Can. J. Plant Sci. 63:753-757.

Wildner, L.P. 2000. Soil cover. p.88-90. IN Manual on Integrated Soil Management and conservation Practices. FAO Land and Water Bulletins 8. IITA and FAO, Rome.

Wolkowski, R.P. 2000. Row-placed fertilizer for maize grown with an in–row crop residue management system in southern Wisconsin. Soil Till. Res. 54: 55-62.

Wortmann, C.S., Xerinda, S.A., Mamo, M. and Shapiro, C.A. 2006. No-till row crop response to starter fertilizer in Eastern Nebraska: I. Irrigated and rainfed corn. Agron. J. 98: 156-162.

In: No-Till Farming
Editor: Earl T. Nardali
ISBN: 978-1-60741-402-5

Chapter 5

SYNERGIES BETWEEN CONVENTIONAL SOIL ORGANIC CARBON, FARM PRODUCTIVITY, SOIL SEQUESTRATION AND SOIL CARBON MARKET RISK IN AUSTRALIA

Mark P. McHenry*
School of Engineering & Energy, Murdoch University, 90 South Street, Murdoch, Western Australia

ABSTRACT

The investment risk of sequestering carbon as soil organic carbon (SOC) is low when compared to the susceptibility of forestry sequestration projects to disease, bushfires and droughts over a 100-year investment timeframe. As it can require decades to create a substantial carbon market asset with the slow natural accrual rates of carbon in soils, farmers will require short-term benefits to their conventional productivity from SOC additions. This chapter explores the additional effects of improving farm SOC levels on conventional farm productivity, soil nutrition, crop yields, soil water use efficiency, crop stability, disease resistance, soil biodiversity and tillage. This work also collates a broad range of management options available to farmers for utilising SOC to assist current productivity, while clarifying the risks and uncertainties of soil carbon stock fluxes, carbon markets and carbon emission policy developments in Australia.

INTRODUCTION

Perceived synergies between climate change mitigation and adaptation can increase the interest in certain technologies and practices that might not increase conventional productivity of the farm. Prospective mitigation and adaptation synergies require a detailed and

* Corresponding author: +61 430 485 306, E-mail address: mpmchenry@gmail.com

comprehensive scientific foundation. In addition, synergistic options require low economic and market risk, must be easily implemented, produce sufficient mitigation and adaptation benefits, and must be financially rewarding. Without reasonable availability of resources alongside scientific, policy and market certainty, it will be unlikely that farmers will be exposing their businesses to unacceptable levels of financial and market risk [1]. Therefore, without sufficient supporting empirical evidence, market-based farm mitigation and adaptation activities will be high risk and will unlikely result in significant sectoral decreases in emissions and reduced climate change vulnerability.

One prospective adaptation and mitigation synergy is to increase the soil organic carbon levels in agricultural lands [2]. Convenietly, the techniques used to increase carbon stocks are common natural resource management (NRM) practices that farmers are likely to be familiar with. Research to date suggests that paddocks with higher levels of SOC show less long-term yield variability, are less sensitive to droughts, maintain good soil fertility, have higher cation exchange capacities, can reduce input costs, enhance trace element availability, reduce soil and wind erosion, leading improved productivity. Therefore, it is sensible to explore adaptation management practices that improve SOC stocks from a conventional perspective, in addition to their potential value in climate change mitigation markets.

Adaptation to a variable climate is not new to Australians, as it has the highest inter-annual natural rainfall variability of all the continents. The principal cause of this natural variability is generally known as the El Niño/Southern Oscillation Index (ENSO), or the Southern Oscillation Index (SOI). Rainfall variability is the principal climatic factor that determines agricultural management in Australia, as yield is closely related to rainfall [3]. Farming systems that have been managed to enhance SOC also exhibit less long-term yield variability and are less sensitive to drought risk than conventionally managed farms [4; 5]. Therefore, increasing farm productive stability and resilience to climate stresses using scientifically verified SOC management practices can reduce some uncertainties for Australian farmers in a natural and anthropogenically variable climate [6; 7; 8; 9].

Agricultural Greenhouse Gas Emissions and Soil C

The natural variability of the Australian climate has lead to the choice of omitting Article 3.4 emissions in the Kyoto Target for Austrlia's National Inventory Report to the United Nations Framework Convention on Climate Change (UNFCCC). Article 3.4 emissions omitted include all native forests under management, revegetation that does not meet forest criteria, and carbon stored in soil and vegetation on both grazing and crop lands [10]. Australian SOC is only reported for the Kyoto target as a component of land-use change activities under Article 3.3. Article 3.4 emissions are excluded because of the risks associated with including large annual fluctuations due to climate dependent emissions (Figure 1).

Unlike most other sectors that emit greenhouse gasses, the agricultural sector emissions are mostly methane and nitrous oxide (Table 1). These emissions are predominantly from enteric fermentation in ruminants, prescribed burning, and microbial activities in soil and water following fertiliser application (Table 2). Australian agricultural soil emission subcategories are comprised of direct, indirect and animal production emissions, and are available at the resolution of the States and the Northern Territory. In the State of Western

Australia (WA), indirect emissions from fertilisers, manures, and burning is the largest subcategory with slightly under half of WA's total agricultural soil emissions. WA's direct agricultural soil emissions from applying fertilisers, manures, nitrogen fixing crops, and crop residues contribute around one-third of the total. The remaining emissions are from animal excretion on paddock. These emission totals, available from the National Greenhouse Gas Inventory (NGGI) for each year from 1990, shows the steady increase in emissions from this subcategory in WA (Figure 2). Reducing agricultural soil emissions in the NGGI requires farmers to be aware of the links between management practices and the resulting direct and indirect emissions.

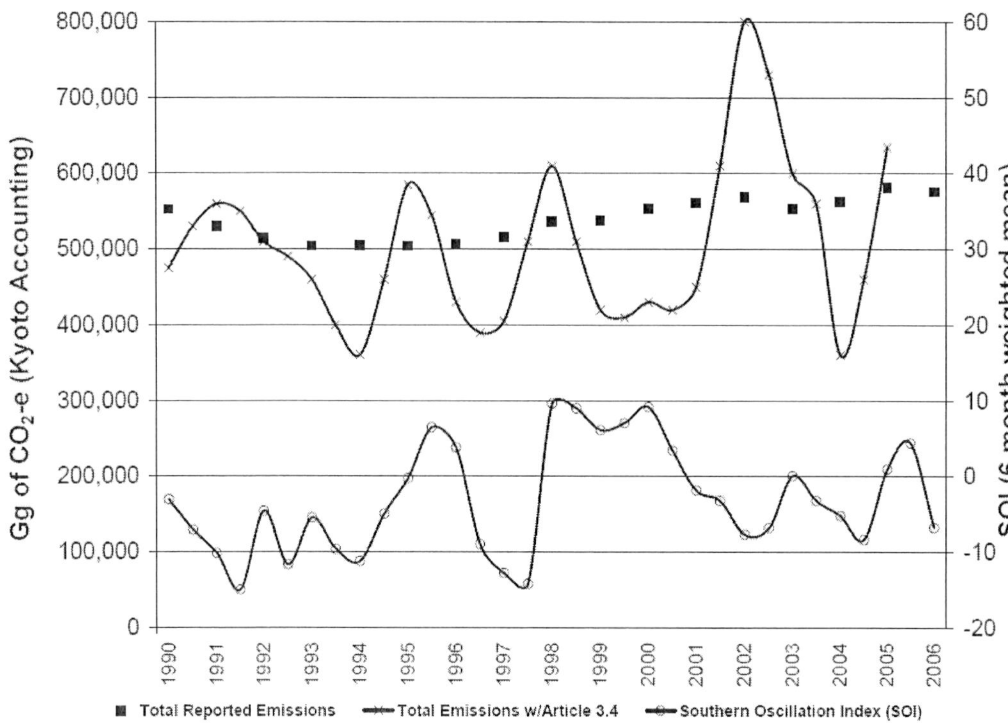

Figure 1. A comparison between Australia's national reported Kyoto emissions, with and without Article 3.4 emissions, and the Southern Oscillation Index (SOI) below.

What is meant by "direct" and "indirect" soil emission categories are sometimes a source of confusion when attempting to reduce overall emissions from farm activities, particularly with the use of fertilisers and manures. Manuring has a number of practical applications in agronomy, but net carbon sequestration is not one of them. This is because more SOC is produced in soils when crop residues are allowed to decompose than when they are fed to animals to produce manure. Nitrogen fertiliser is also commonly recommended to increase SOC. While this is true for local-scale farming in terms of reducing farm greenhouse gas emissions with the increase of SOC in the 0-30 cm of topsoil layers, the amount sequestered is insufficient to balance global emissions produced during fertiliser manufacture [11]. While pockets of practical and technical difficulties to successful soil sequestration remain, there is a need for parallel work on reducing political and market barriers to farmers using soils to generate offsets under emission trading schemes.

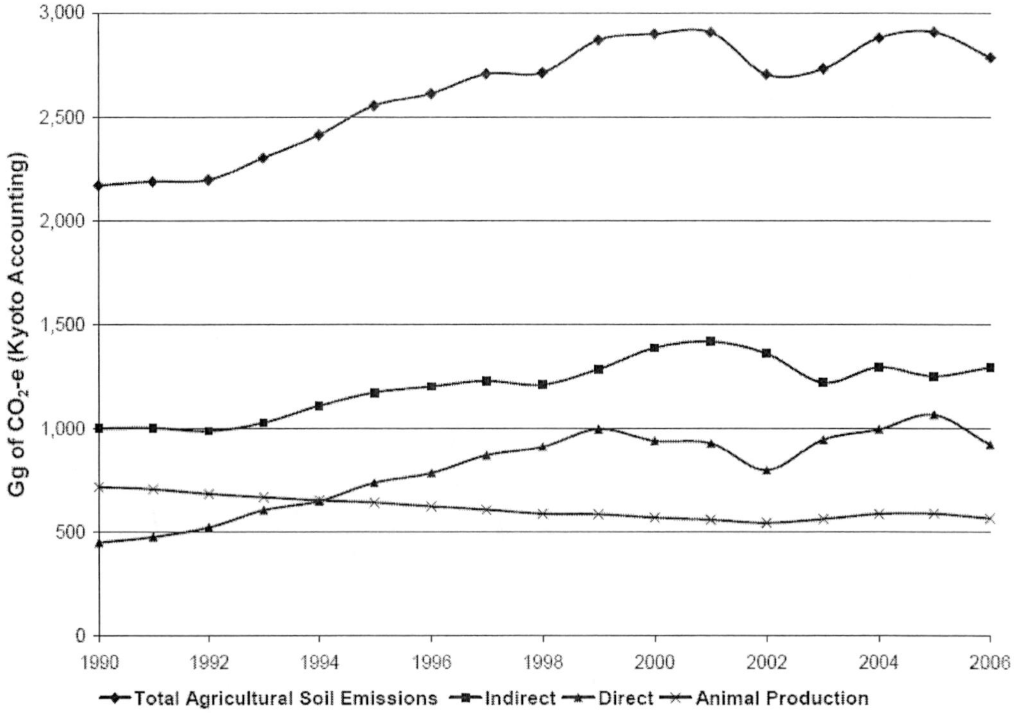

Figure 2. Western Australian agricultural soil emissions from 1990 to 2006.

Table 1. Australian 2006 sectoral net greenhouse gas emissions (Mt, Kyoto accounting).

Sector & subsector	CO_2	CH_4	N_2O	HFCs/PFCs/SF_6	CO_2-e	%
Stationary energy	285.30	1.10	1.00	-	287.40	**49.90**
Transport	76.80	0.60	1.70	-	79.10	**13.73**
Fugitive emissions from fuel	5.80	28.70	0.02	-	34.50	**5.99**
Industrial processes	22.60	0.10	0.02	5.80	28.40	**4.93**
Solvent and other product use[(a)]	-	-	IE	-	IE	**-**
Waste	0.03	16.00	0.60	-	16.60	**2.88**
Land use, land use change & forestry	37.40	2.10	0.60	-	40.00	**6.94**
Agriculture (total)	-	69.80	20.32	-	90.10	**15.60**
Ag. subsector (Enteric fermentation)	-	59.30	-	-	59.30	10.30
Ag. subsector (Manure management)	-	2.00	1.60	-	3.60	0.60
Ag. subsector (Rice cultivation)	-	0.30	-	-	0.30	0.05
Ag. subsector (Agricultural soils)	-	-	15.20	-	15.20	2.60
Ag. subsector (Prescribed burning of savannas)	-	8.10	3.40	-	11.50	2.00
Ag. subsector (Field burning of ag. residues)	-	0.20	0.10	-	0.30	0.10
National Total Net Emissions[(b)]	**427.80**	**118.30**	**24.20**	**5.80**	**576.00**	

(a) Emissions are included in Industrial processes for reasons of confidentiality. (b) The net credits from land use, land use change & forestry should only be included in this account during the first commitment period. They are included in the 2006 Inventory as a guide for the Kyoto target, which is estimated to be 597 Mt each year over the period. IE = Included elsewhere. - = Not applicable. Sector percentages and totals will not add due to rounding.

Table 2. Western Australian 2006 agricultural soil emission subcategory totals (Gg, Kyoto accounting).

Indirect	Atmospheric Deposition	Fertilisers		125.33
		Manures	Cattle	112.66
			Sheep	162.98
			Others	19.51
		Others	Burning of agricultural residues	5.57
			Prescribed burning of savannas	356.71
	Nitrogen Leaching and Run-off	Fertilisers		205.76
		Manures	Cattle	141.75
			Sheep	155.74
			Others	7.74
			Indirect Subtotal	**1293.75**
Direct	Synthetic Fertilisers			507.1
	Animal Waste Applied to Soils			48.98
	Nitrogen Fixing Crops	Legume Pastures		144.76
		Pulses		18.46
	Crop Residues	Cereals		180.79
		Pulses		12.37
		Other Crops		11.14
			Direct Subtotal	**923.60**
Animal Production	N Excretion on Pasture & Paddock	Urine		405.18
		Faeces		161.97
			Animal Production Subtotal	**567.15**
WA Agricultural Soil Total Emissions (Gg of CO_2-e)				**2784.50**

AGRICULTURAL SOC OFFSET MARKETS, POLICIES, AND RISKS

Various government documents related to Australia's emission trading scheme, the Carbon Pollution Reduction Scheme (CPRS) explicitly advocate further research into both measuring and retaining carbon in soils to provide comprehensive and scientifically accurate national carbon accounting and monitoring methodologies [10]. Developing standardised methods to quantify SOC sequestration for specific land and soil-use changes streamline offset certification and verification. Unfortunately, the use of standardised accounting methodologies without sufficient model verification may result in inadequate estimates of the actual carbon sequestered in soils. This inadequacy is likely to result in discounting of soil carbon offsets in trading markets to the point where Australian SOC offset projects will not occur.

Australian SOC offset products are ineligible to be traded on the Chicago Climate Exchange (CCX), as the CCX does not have an existing protocol that allows Australian soil offsets to generate tradable Carbon Financial Instrument (CFI) contracts. Trading of Australian SOC offset CFI products will require researching carbon uptake from soils from various soil cultivation methods to a level similar to the CCXs existing North American Soils Protocol. As Australia does not include Article 3.4 emissions in the Kyoto reporting, there remains a possibility of eventually trading Australian soil-derived CFI's on the CCX, if a sound sampling methodology and a protocol is approved. Greenhouse Friendly™ (GF), a Australian Commonwealth Government initiative, was launched in 2001 to provide a level of consumer confidence in offset markets by certifying voluntary offset products and services. This was undertaken through an independent verification process of individual abatement projects. To date, there are no existing or prospective projects approved by GF that achieve abatement through the sequestration of carbon in soils. This is primarily due to problems associated with monitoring and verification procedures. If a reliable monitoring procedure is developed that meet independent verification requirements, the GF has the capacity to approve projects specialising in SOC sequestration using technologies that enhance soil carbon levels, coupled with forestry projects.

The Australian National Carbon Accounting System (NCAS) and the "Full Carbon Accounting Model" (FullCAM), accounts for carbon stock changes based on land use management. FullCAM estimates and predicts all biomass, litter, and soil carbon pools in forest and agricultural systems and for changes in major greenhouse gases, nitrogen cycling, and human-induced land use practices. The soil carbon modelling inputs include the soil clay content, temperature, soil moisture content, plant and manure inputs, plant cover, in addition to microbes, and multiple active soil pools [12]. If accredited verifiers for forestry projects use FullCAM to estimate the encapsulation of active pools to the inert soil carbon pool, the results may require an additional verification study to ensure the modelling concurs with what is occurring on-site. Therefore, despite some very detailed soil research and useful biomass-soil carbon flow modelling software, it is likely that both passive and active SOC sequestration to forestry projects will still require verification by direct soil sampling.

The Australian Department of Climate Change acknowledges there is much research and development required before a comprehensive modelled account of SOC is achieved for Australian forestry projects, soil management changes, and direct biomass application to soils. For practical reasons, national SOC stock accounts are modelled to reduce the costs of periodicly measuring soils directly over vast distances. Such models still require substantial amounts of input data, including soil types, initial carbon content, clay component, land uses, climate, and residue inputs, in addition to calibration and verification. The NCAS soil carbon model validation results are generally very good. However, difficulties remain in reliably modelling changes in coarse woody debris and biomass litter turnover from input, transfers and loss information, due the complexity of its dynamic nature in relation to soils [10]. At this stage, the direct sampling of SOC sequestration is the least-risk option for project developers of enhanced SOC offsets.

Future developments in accounting and verification of direct soil sampling of passive and actively enhanced SOC pools will enable the creation of Verified Emission Reductions (VERs) under a methodology similar to the GF program. However, if Australian agriculture is captured in the CPRS, then it will severely limit the scope to create agricultural offsets and the incentives for developing offset methodologies [10]. While the agricultural sector will not be included in the CPRS until at least 2015 (if at all), the uncertainty of how such a policy will capture the variety of farm activities and emissions has strong parallels with the current issues faced by the waste sector. In the future, even without agricultural CPRS capture, large piggery and dairy waste recovery activities, including biogas renewable energy generation options, may have a similar liability to the refuse-waste recovery sector under a CPRS. Agriculturalists would be well advised to follow the developments of how waste recovery facilities handle the introduction of the CPRS, as the two sectoral emission profiles, and resources they manage are largely biological.

Strong parallels also exist between the external risks faced by "early adopters" of mitigation options in both the waste and agricultural sectors. The greatest risk being Australian Government policy changes relating to CPRS capture. The Australian and New Zealand Standard Industrial Classification (ANZSIC) titles and codes are used to collect and publish statistical information in the two countries. The ANZSIC codes will be used to determine businesses that are captured under the Australian CPRS, and currently large domestic biomass waste-to-compost manufacturers will be included. This means that it's possible for some businesses to have a liability under the CPRS, and be accredited to generate offsets under programs such as GF. One example in WA is the South Metropolitan Regional Council's Regional Resource Recovery Centre (RRRC). The RRRC is a GF approved integrated waste processing facility abatement project. The RRRC produces significant useful biomass as it diverts and recovers 85 percent of domestic waste from landfill to produce compost, mulch or is separated and delivered to manufacturers for recycling. It's processing capacity includes 169,000 t for various biomass inputs in addition to 30,000 t of materials recovery for co-mingled recyclables [13]. Existing waste recovery abatement activities that generate VERs might become unprofitable if they attract liabilities under the CPRS. Without being able to offset their own financial liabilities by generating undiscounted SOC VERs, waste recovery projects might be the unfortunate loser under some CPRS capture scenarios. The RRRC currently generates 80,000 VERs annually from their waste recovery activities and sell to customers that voluntarily offset their emissions. These customers include Origin Energy, Virgin Blue, BP, Synergy, the Carbon Reduction Institute, the Australian Climate Exchange and others. The financial impact for entities unable to sell these offsets if their operations exceed 25,000 tCO_2-e in a financial year under a future CPRS, will also attract the new administration costs associated with CPRS implementation.

CPRS administration costs are likely to be similar to the costs of independently assessing VERs under the GF program. To become an approved generator of VERs, the original RRRC GF project design document for the RRRC included a ballpark AUD$20,000 consultancy, with ongoing annual reporting requirements costing around AUD$15,000 for the independent verification process. The RRRC does not currently receive VERs for the additional SOC that any biomass compost sequesters, only the avoided emissions from not sending the waste to landfill. If waste facilities are captured under the CPRS, the emissions and accounting activities can quickly become complex. Much difficulty results from the variety of waste processing options available, and the choices of at which point in the processing cycle the emissions or offsets may be acquitted against liabilities. As net life-cycle emissions from composting abates around 50 kg of CO_2-e per tonne of composted yard trimmings and food waste, the additional soil sequestration combined with the improved tree growth seems like a win-win situation for both forestry projects and biomass waste-to-compost facilities wishing to increase abatement. However, who owns the carbon rights to the compost may become an issue. Sidestepping these problems may require alternatives to composting biomass, such as directly combusting the waste. This combustion also reduces net CO_2-e life-cycle emissions of yard trimmings and food waste by 60 and 50 kg of CO_2-e per tone, respectively [14]. If pyrolysis technologies were used instead of complete combustion options, the waste biomass could be converted to carbon-dense stable forms of charcoal for use as a valuable soil conditioner. Using pyrolysis technologies, around half of the carbon in the biomass may be sequestered while generating roughly two-thirds of the bioenergy produced when the biomass is completely combusted [15]. The question of when emissions are accounted for under the CPRS, and/or who owns the right to the carbon or holding a liability, remains open for now.

If agriculture is likely to be captured after 2015, then there will be little incentive to develop offset methodologies and institutional infrastructure to support an offset market for such a short period. The final decision of whether agriculture is included in the CPRS (expected in 2013), will also determine the existence of a formal agricultural offset scheme [10]. Even without agricultural CPRS capture, sequestering carbon in soils does not come without market risk. For example, if an offset business sold VERs in the beginning of 2008 that are required to be maintained for 100 years, and this offset was either removed or was proven to be non-existent, then the business would be liable to either replace the offset, or purchase VERs to cover any shortfall. With the expected increases in carbon prices over-time, the financial risk of selling SOC VERs prematurely at discounted market prices, using unverified models are high. For example, if the RRRC was required to repurchase their 2007 VERs traded on the Australian Climate Exchange (ACX) only one year after selling them, they would need to pay an additional 30 percent (excluding transaction costs), due to the rising market value of their VERs between 2007 and 2008 (Figure 3). With the price of carbon generally expected to increase over time, the financial risk of prematurely selling VERs that require long-term maintenance is a risk that may be preventing mitigation projects from being developed further.

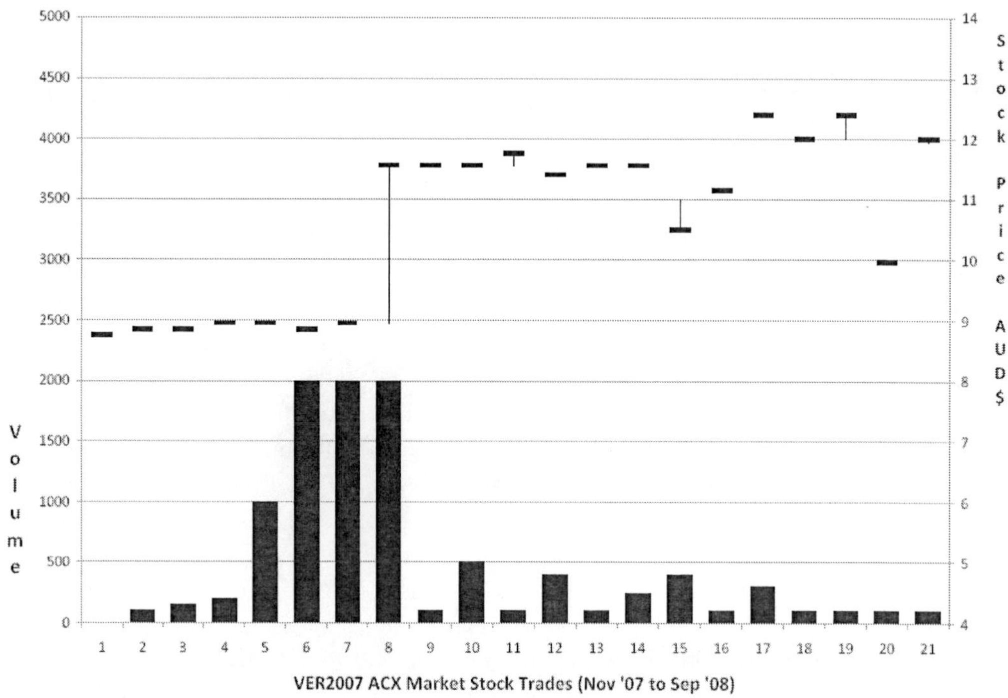

Figure 3. Total VER2007 product trade price and volume on the Australian Climate Exchange.

SOC Mangement to Decrease Liabilities and Increase Productivity

The global decline in SOC as a result of deforestation and arable cropping have made significant contributions to increased levels of atmospheric carbon dioxide [16], as cited in [17]. As a result of climate change mitigation policies, there is renewed interest in the role of agricultural soils in reducing emissions and enhancing biological carbon sequestration. Returning SOC back into the soil gives farmers an indication of how it may effect primary productivity, while monitoring future SOC carbon market developments. Farms already using economically efficient practices, may receive lower financial returns if they prematurely attempt to sequester carbon. For farmers to avoid this income opportunity cost they should understand their farms soil emissions, it's sequestration potential, and the appropriate management practices that both enhance productivity and reduce the risk of losing a carbon asset [18].

SOC, Farm Nutrient Availability and Water Use Efficiency

Managing pasture soils to increase carbon levels increase cation exchange capacities and enhance availability of trace elements [19; 20]. The ability of soils to retain nutrients in cation forms available to plants, is known to increase in proportion with soil organic matter [21]. High nutrient use efficiencies occur when a farmer can obtain large amounts of plant growth from soils with low nutrient levels [9]. As a general rule, when carbon in soils is lost, nitrogen is often also lost, and when carbon is gained, nitrogen is also gained [22]. Increasing the ability of pastures to use nitrogen and nutrients can reduce the expenditures required for fertilisers and soil conditioners.

The balance of organic matter production and its decomposition by soil organisms determines SOC stock levels. Both of these processes are strongly dependent on physical, chemical and biological factors, such as climate, nutrient availability, plant growth patterns, and soil water status [23; 24]. In some circumstances, the build up of repellent organic matter in sandy soils can slow water penetration into dry soils. Water repellency is common in annual cropping systems, but has also been reported to increase under long-term lucerne and veldt grass stands [25], as cited by [26]. Due to the lack of water in the soil in early summer (December), productivity declines in the southwest jarrah forest regions of WA, while maximum productivity occurs in late winter to early spring from winter rains [27]. Reducing evaporation late in the growing season will conserve soil water to improve plant water use efficiency and extend the season [28]. Increasing SOC stocks can lower soil bulk densities to allow greater water penetration to increase pasture productivity [19; 20]. Preserving water in the soil up to harvest can control soil loss, as higher moisture at this time reduces soil erosion [29]. Maintaining soil nutrients, reducing fallow times, and tillage will also enhance water storage and water use efficiency [30; 31]. Composting and green manures also increase the SOC stocks while improving soil structure, reducing erosion, evaporation, and providing some protection against raindrop impact. Composts and mulches can modify temperature and moisture, although the effects are linked to the quality and quantity of the biomass residues. In some circumstances mulches are even an alternative to herbicide use [32].

Pasture Types and Their Effect on C Soil Stocks

Maintaining plant cover throughout the year is probably the most important factor in reducing erosion by both water and wind. As perennials are not required to germinate each year, their deeply penetrating roots improve soil structure, and create biopores that enhance drainage to reduce water repellence [26]. Deep-rooted perennials can use water when annual pastures are dead, recover leached nitrate, reduce acidification, provide a soil cover, and a root mass to restrict erosion through summer [26; 33]. Ferdowsian and Greenham (1992) found that annual pastures in the southwest of WA had few roots penetrating deeper than half a metre. However, perrenail pastures such as phalaris and yorkshire fog roots extended to one metre, and kikuyu roots had the largest root mass that extended below one metre. When comparing perennial pasture types, the deeper rooting habit of phalaris and tall fescue offer more scope for overcoming degraded soils than does cocksfoot and perennial ryegrass, while kikuyu remains green and productive throughout summer [26]. Deeper and denser root systems of perennial forage species are generally more competitive with windbreak trees in comparison to annual crops, which is a very important benefit in integrated systems [34].

Integrating trees and shrubs into pastures improve the soil nutrient cycle. Deeper root systems cycle nutrients from sub-layers of soil into the topsoil to a greater extent than annuals [35]. Shading of the soil by perennial pastures also reduce the rate of organic matter decomposition, which increases SOC stocks and associated benefits [35]. A study by Resh, Binkley and Parrotta (2002) on nitrogen fixing perennial trees showed that the levels of SOC under the nitrogen fixers was between 10 and 90 percent greater than the soil under Eucalyptus trees. The greater SOC levels under the nitrogen fixing trees resulted from both the greater retention of old SOC and the greater accretion of new SOC [36].

Combinations of perennial grasses with trees reduce runoff, with kikuyu and subterranean clover often used for this purpose, as both are shade tolerant [34]. Nitrate accumulates when subterranean clover plants break down in summer and leach from the soil profile following autumn rain on annual pastures. This can occur before annuals can develop an actively growing root system that captures significant levels of nitrate [37]. Perennial trees and pastures can improve nitrogen use efficiency to a greater extent than smaller annuals in the early season. Nutrient topdressing in autumn before an annual pasture develops active roots systems, or sufficient plant cover, often results in both fertiliser and topsoil loss, especially with heavy opening rains. In Albany on the south coast of WA, up to 60 percent of the applied phosphorous was lost in particulate form during one autumn storm [38] as cited in [26].

One example of a detrimental pasture management practice is the permanent removal of crop residues over summer. The removal of biomass is inversely related to SOC accumulation, as the remaining biomass eventually decomposes and contributes to SOC and nutrient levels [39]. Overgrazing can also limit the amount of plant biomass growth and surface cover, resulting in less fodder availability over the growing season and reduced SOC accrual. However, leaving some crop residues in place by reducing summer grazing intensity and stock numbers will require careful consideration by farmers. At this time, more research is required to link specific combinations of grazing management practices to increased feed productivity and soil carbon stocks in particular climates [40]. The main constraints to the adoption of perennial pastures are the perceived costs and their poor persistence under continuous grazing, especially by sheep [26]. Nonetheless, it is likely that temperate perennial grasses, in conjunction with legumes, could provide a sustainable pasture base in southern Australia, only if animal grazing is well managed [26].

The inclusion of perennial forages in rotations increase soil carbon levels relative to rotations with annual crops alone [30]. Continuous and diverse pasture growth and grazing most closely simulates perennial systems that lead to minimal nutrient loss and the maximum accumulation of biological activity [31]. Soil carbon levels also tend to increase with crop rotations, continuous cropping, cover crops, and mixed cropping [30; 31; 32]. Rotations of crops are an effective management strategy for annual monocultures, while cover cropping assist orchards and vineyards to achieve greater levels of biodiversity and stability. Orchard cover crops are commonly and successfully used to fix nitrogen, modify the microclimate, provide a habitat for beneficial insects, encourage soil biology, and add SOC [19].

Soil Biodiversity, Crop Stablity, and Disease Resistance

Most of the biodiversity of agricultural systems resides in the soil, but the relative understanding of this scale of biodiversity is poor [41]. Biodiversity comprises the "planned" biodiversity, such as the crops and/or livestock the farmer wishes to produce, as well as the "unplanned" biodiversity, or all the other biota in the system. The value of the aboveground planned biodiversity is more obvious to the observer than the intrinsic value of the soil biodiversity below our feet. Furthermore, most biological organisms are viewed and treated as harmful (pathogens, pests and weeds), although many are in fact beneficial (pollinating insects and pest predators). They should therefore be managed to either increase or decrease their populations, based on their individual merit. Unfortunately, the management of soil biodiversity is not as simple as choosing livestock and crop varieties [9].

Increased soil biota assists biological immobilisation of inorganic nitrogen, which also aids in retaining nitrogen, adsorbs dissolved ammonium, nitrate, phosphate, and water repellent organic pollutants [15; 42; 43; 44; 45]. Food-web interactions between soil biota and plant roots also have a large impact on SOC fixation, crop quality, pest predation, populations of beneficial organisms that cycle nutrients, and the incidence of soil-borne plant and animal pests/diseases [9]. Resistance against outbreaks of pests and diseases, and resilience from disturbance is of particular importance to agricultural cropping [46]. High total microbial biomass and a strong competition for carbon and nutrients, leads to the suppression of specific pathogens by particular competitors operating in a background of general competition [9; 19]. Increasing soil biodiversity for crop resilience against stress and disturbance can be regarded as a form of natural protective insurance [9]. Stability is often defined as the ability to recover from stress or disturbance, whereas resilience refers to the rate with which populations recover from stress or disturbance [6; 7; 9]. A short-term transient event is known as a "disturbance" where a more persistent pressure, such as a heavy metal contamination that continuously effect organisms, is referred to as a "stress" [9; 47].

Experiments by Griffiths *et al.* (2000) on the relationship between stresses and disturbances on biodiversity and ecosystem function showed that high and low soil biodiversities recover from stresses at different rates. This research involved creating stresses such as temperature extremes and toxic chemicals. High soil biodiversities were associated with a greater resumption of normal ecological functions to pre-stress levels after the loss of some populations. These experimental results showed that even though stressed soils exhibit a significant reduction in biodiversity, the impact on the ecological functions they performed was minor. Interestingly, when these same stressed soils underwent another disturbance event were compared with similar unstressed soils, both undergoing another disturbance event, the ecological functions were more severely effected, and for a longer time in the stressed groups, than the unstressed groups [6; 9]. Therefore, the response of the microbial communities to stressors depends on the history of the organisms, and the type of stressor applied [7; 48; 49]. Thua, biodiversity does not necessarily confer ecological function stability, but does result in a higher resilience, and an ability to recover quickly from disturbances [7; 9].

Soil biodiversity is more meaningful to farmers when integrated with aboveground productivity. One of the most well known uses of soil organisms in agriculture is to fix nitrogen from the air into the soils. Biological nitrogen fixation is provided by soil biota in leguminous plants, and may fix more than 100 kg of nitrogen per hectare per year. Total annual contribution of nitrogen by micro-organisms in both agricultural and natural ecosystems has been estimated at between 140 to 170 million tonnes annually, and valued at US$ 90 billion per annum. Other useful soil biota functions and their estimated annual values include: recycling of organic wastes, with bio-remediation of polluted soils and water (US$ 121 billion); the control of pests in agricultural systems (US$ 160 billion); the use of various wild insects, plant roots and mushrooms as food for humans ($US 180 billion), and; the pollination of plants by insects that spend a critical stage of their life within the soil (US$ 200 billion) [9].

Biological Management in Pasture and Soils

Farm management options that improve soil biological processes, increase biodiversity and sustain a viable community of soil biota include: increasing the amount and/or quality of organic residues; using biological control for pests and diseases; inoculating for beneficial soil organisms, (such as disease antagonists, microsymbionts, rhizobacteria); minimising soil disturbance, pesticide use, irrigation, and artificial fertilisers, and; using earthworms for disease control and soil fertility [9]. Under some conditions fertiliser applications may decrease SOC and soil microbial populations, but the beneficial effects of fertilisers generally offset any adverse effects as increased plant residues increase microbial activity [32]. Microbial activity is highest in organically fertilised agricultural soils and manure applications tend to increase the earthworm biomass in cropped soils, while reduced tillage creates a more stable environment [19]. Tillage disrupts soil aggregates, compacts soils, and disturbs plant and animal communities that contribute to aggregation, and leads to lower microbial activity [32].

Herbicide residue can also negatively effect soil biodiversity by changing the soil microclimate [19]. A two-year study by Caldeira *et al.* (2001) on soil moisture, water availability, and water use by plants in various soil biodiversities found significantly higher total biomass in species-rich areas alongside a reduction in water consumption, when compared to monocultures. The results suggested that the species-rich plant communities had more water available in the upper soil where the roots were concentrated. In the majority of seasons, soil polycultures significantly increased the uptake efficiency of nitrogen and phosphorus relative to monocultures, which some research suggests may be due to soil biota and soil biodiversity [9].

A microcosm study by Heemsbergen *et al.* (2004) revealed that it may not be general biodiversity that enhances soil nutrient supply, but the functional and evolutionary dissimilarity of soil organism species [9; 50]. Increased soil mycorrhizal fungal diversity is positively correlated with increased water uptake efficiency. This effect may be indirectly related to mycorrhiza-mediated effects on soil structure [9]. Auge (2004) showed that the effect of mycorrhizal fungi on stomatal conductance was approximately equally due to root and soil colonisation. Therefore, even plants without the fungi colonies benefit from mycorrhizal fungi water uptake, and the increased soil moisture availability for sustained periods of growth, and as a consequence, improved SOC fixation [9; 51].

In the case of agroecosystems and soils, the objective is to maintain key functional groups. Therefore, it is important to know if key species exist, what they are, and how many are necessary to sustain soil processes and primary crop productivity. Testing soils for biological diversity may become a cheap and widely available tool to accurately monitor system changes, as organism populations reflect specific characteristics of soils and SOC stocks [50]. However, there is significantly more research required to validate biological assessment and management practices suited to specific agro-ecosystems in Australian soils and climates [9].

Tillage and Carbon Stocks

Conventional tillage consists of disking, chiselling, ridging and residue incorporation into soil, while no-tillage methods leave crop residues on the soil surface [52]. Conventional tillage contributes to soil disturbance and is associated with more traditional cropping practices [53]. Losses of SOC of as much as 50 percent in surface soils (up to 20 cm depth) have been observed after cultivation for 30-50 years, with reductions averaging around 30 percent for the top one metre [54]. The effect of tillage in higher rainfall areas is greater than in areas with lower rainfall, as water is less of a limiting factor in terms of plant growth [40]. For mean annual temperatures between 12.8 and 17.4 degrees Celsius, where average rainfall is higher than 650 mm, soil management practices have a definite effect on SOC density up to 30 cm, with higher densities corresponding to lower tillage and stubble retention [40; 53]. A SOC density comparison between Australian farms using minimum disturbance methods have been found to be around 57 tonnes per hectare (t/ha), while nearby farms using conventional tillage practices exhibited 43 t/ha, both up to 30 cm of soil depth [40].

Tillage, in addition to mixing and stirring of soils, breaks up aggregates and exposes surfaces otherwise inaccessible to decomposers [54]. Conservation tillage practices are associated with reduced, or no soil disturbance, and include practices ranging from "no-tillage/direct drill" to "reduced tillage/minimum tillage" [40]. The use of no-till can increase SOC stocks in annual crop systems, provided that crop production is not adversely effected [30]. No-tillage or direct drill systems retain stubble for the maximum length of time prior to sowing a new crop., ground disturbance is kept to a minimum during sowing, and seedbeds are not tilled prior to sowing. Permanent beds, raised beds, controlled traffic and precision agriculture are also grouped under the no-tillage classification [40].

Reduced tillage aims to minimise soil disturbance, while achieving a viable seedbed. As with no-tillage systems, weeds are controlled by herbicide and grazing and crop stubble is often burnt and/or incorporated into the soil. Farmers using reduce tillage often utilise implements that minimise the area, depth, and extent of soil disturbance on the soil structure

[40]. A study by Wright, Dou and Hons (2007), on SOC and nitrogen after 20 years of conservation tillage, found that the use of conventional tillage did not increase the SOC or total nitrogen in subsurface soils to levels greater than no tillage. The study suggested that burial or incorporation of residues was ineffective at increasing SOC or nitrogen accumulation over the 20-year period [52].

Wright *et al.* (2007) found conventional tillage reduced dissolved organic carbon, SOC, and total nitrogen in surface soils (0-5 cm) under continuous wheat crops, and in subsurface soil depths down to 55 cm for more intensive cropping sequences. No-tillage increased SOC, dissolved organic carbon and total nitrogen, compared to conventional tillage by 28, 18 and 33 percent, respectively. High intensity non-monoculture cropping sequences, coupled with no-tillage resulted in the highest SOC stocks in subsurface soils, demonstrating the importance of subsurface carbon and nitrogen for potential carbon sequestration. For more intensive cropping sequences under no-tillage, decreasing fallow periods and increasing crop residue production increased SOC and nitrogen levels to the 55 cm depth [52]. Therefore, long-term conservation tillage may include protection of SOC and total nitrogen in subsurface soils, and should assist in maintaining farm carbon stocks [52; 53].

A report on the NCAS by Valzano, Murphy and Koen (2005), focussed on the impact of tillage on changes in SOC density in Australia. They found that changing from moderate to low intensity grazing, when in combination with any cropping tillage practice, would usually lead to a loss of SOC. The report stated that tillage practices alone are insufficient to predict likely soil carbon densities. This is because the set of management practices included in "low tillage" or "no-tillage" umbrellas range widely, and in terms of the timing of various management practices in a season [40]. The NCAS report found the differential between conventional tillage practices and reduced tillage on paddock SOC densities was up to 10 t/ha. Interestingly, this difference was not apparent in the top 10 cm of soil. The report found differences of up to 25 percent between conventional tillage practices and conservation tillage, with the greatest SOC losses occurring with higher levels of soil disturbance, in combination with stubble burning. Stubble burning was a traditional method used in late summer after harvest. At present, if stubble is burnt, it is now burnt just prior to sowing when fire restrictions are lifted, which is usually around March in the southern half of WA [40]. This reduces the likelihood of bushfires, and reduced the times that soils are exposed to ambient conditions and erosion.

Crop stubble may be burnt, grazed, or incorporated in either conservation or conventional tillage systems. Stubble incorporation involves the use of tillage implements to bury remnant plant residues, and has traditionally been used to return organic matter to the soil, and to protect from erosion. The NCAS report suggested that incorporation can transfer pathogens to other paddocks, offer less protection than stubble retention practices, and can destroy soil structure and porosity [40]. Stubble retention involves leaving crop residues at the soil surface, which can be grazed just prior to sowing another crop. The report suggested that stubble retention is the best stubble management method, as it protects the soil surface from erosion and retains carbon at the soil surface [40; 53].

Conclusion

Balancing farm yields and decreasing the required level of conventional farm inputs can decrease farmer uncertainty, especially in periods of rising costs, labour shortages and climate variability [55]. The cultivation of an effective soil ecology can reduce fertiliser requirements, improve water use efficiency and prevent plant and nutritional deficiencies [9; 19; 20; 21; 30; 31. The effective management of SOC can positively influence soil fertility and soil physical properties, such as aggregate stability and water holding capacity, to improve the farm ecosystem [55]. Despite these benefits, SOC is not routinely included in basic soil testing regimes in Australian agriculture, nor does it receive the recognition it deserves as a means for greenhouse gas mitigation.

Carbon sequestration in agricultural soils also has the potential to create an economic commodity for farmers, while improving conventional productivity and providing ecological benefits [1]. However, even using promising pasture combinations, tillage practices and grazing levels, SOC densities can remain low given certain combinations of climate, soil type and past management practices [40]. More research and communication of farm soil management practices that foster SOC stocks for specific agricultural systems, and combinations of soils and climates, are required to provide sufficient certainty to soil carbon markets in Australia. This primary soil research requires continued parallel policy and accounting methodology development to provide sufficient security to both farmers and investors, to ensure long-term benefits from soil carbon markets.

Acknowledgments

Thanks to Tim Youé, the Manager of Business Development at the Southern Metropolitan Regional Council for offering his expertise and insights for this chapter. Also, a big thank you to Julia for persisting in proof-reading drafts of her favourite subject ;)

References

[1] Klein, RJT; Huq, S; Denton, F; Downing, TE; Richels, RG; Robinson, JB; Toth, FL. *Interrelationships between adaptation and mitigation*; Cambridge University Press: Cambridge, *United Kingdom*, 2007.

[2] International Panel on Climate Change *Summary for Policymakers. In: Climate Change*, 2007*: Synthesis Report*, 2007.

[3] Risbey, J; Kandlikar, M; Dowlatabadi, H; Graetz, D. Scale, context, and decision making in agricultural adaptation to climate variability and change. *Mitigation and Adaptation Strategies for Global Change*, 1999, 4, 137-165.

[4] Lotter, DW; Seidel, R; Liebhardt, W; Dou, F. The performance of organic and conventional cropping systems in an extreme climate year. *American Journal of Alternative Agriculture*, 2003, 18, 156-154.

[5] Lugato, E; Paustian, K; Giardini, L. Modelling soil organic carbon dynamics in two long-term experiments of north-eastern Italy. *Agriculture, Ecosystems and Environment*, 2007, 120, 423-432.

[6] Griffiths, BS; Ritz, K; Bardgett, RD; Cook, R; Christensen, S; Ekelund, F; Sorensen, S J; Baath, E; Bloem, J; de Ruiter, PC; Dolfing, J; Nicolardot, B. Ecosystem response of pasture soil communities to fumigation-induced microbial diversity reductions: an examination of the biodiversity - ecosystem function relationship. *Oikos*, 2000, 90, 279-294.

[7] Tobor-Kaplon, MA; Bloem, J; Romkens, PFAM; de Ruiter, PC. Functional stability of microbial communities in contaminated soils. *Oikos*, 2005, 111(1), 119-129.

[8] Harle, KJ; Howden, SM; Hunt, LP; Dunlop, M. The potential impact of climate change on the Australian wool industry by 2030. *Agricultural Systems*, 2006, 93, 61-89.

[9] Brussaard, L; de Ruiter, PC; Brown, GG. Soil biodiversity for agricultural sustainability. *Agriculture Ecosystems and Environment*, 2007, 121(3), 233-244.

[10] Department of Climate Change *Carbon pollution reduction scheme green paper*; Commonwealth of Australia: *Canberra*, 2008.

[11] Schlesinger, WH. Carbon sequestration in soils: some cautions amidst optimism. *Agriculture Ecosystems and Environment*, 2000, 82, 121-127.

[12] Australian Greenhouse Office *National carbon accounting system: the FullCAM carbon accounting model version 3.0 user manual*; Commonwealth of Australia: Canberra, 2005.

[13] South Metropolitan Regional Council Regional Resource Recovery Project: a brief summary of technical facts. http://www.smrc.com.au/projects/rrrc/rrrc-tech.cfm (20 October),

[14] United States Environmental Protection Agency Solid waste management and greenhouse gases: a life-cycle assessment of emissions and sinks. 3rd Edition. http://www.epa.gov/climatechange/wycd/waste/downloads/fullreport.pdf (30 October),

[15] Lehmann, J; Gaunt, J; Rondon, M. Bio-char sequestration in terrestrial ecosystems. *Mitigation and Adaptation Strategies for Global Change*, 2006, 11, 315-419.

[16] Post, WM; Peng, TH; Emanuel, WR; King, AW; Dale, VH; De Angelis, DL. The global carbon cycle. *American Scientist*, 1990, 78, 310-326.

[17] Grace, PR; Post, WM; Hennesey, K. The potential impact of climate change on Australia's soil organic carbon resources. *Carbon Balance Management*, 2006, 1(14), 1-10.

[18] Antle, JM; Stoorvogel, JJ; Valdivia, RO. Assessing the economic impacts of agricultural carbon sequestration: Terraces and agroforestry in the Peruvian Andes. *Agriculture, Ecosystems and Environment*, 2007, 122(4), 435-445.

[19] Altieri, M. A. The ecological role of biodiversity in agroecosystmes. *Agriculture Ecosystems and Environment*, 1999, 74, 19-31.

[20] Jones, C. Australian soil carbon accreditation scheme (ASCAS). In *Managing the Carbon Cycle Workshop*, Amazing Carbon Katanning, 2007.

[21] Lehmann, J. Bio-energy in the black. *Frontiers in Ecology and the Environment*, 2007, 5(7), 381-387.

[22] Murty, D; Kirschbaum, MU; McMurtrie, RE; McGilvray, H. Does conversion of forest to agricultural land change soil carbon and nitrogen? A review of the literature. *Global Change Biology*, 2002, 8(2), 105-123.

[23] Post, WM; Izaurralde, RC; Mann, LK; Bliss, N. Monitoring and verifying changes of organic carbon in soil. *Climatic Change*, 2001, 51(1), 73-99.

[24] Kirschbaum, MUF. Will changes in soil organic carbon act as a positive or negative feedback on global warming? *Biogeochemistry*, 2000, 48(1), 21-51.

[25] King, PM. Farming water repellent sands. In *Lucerne logic - the role of dryland lucerne in Western Australia*, Jerramungup Soil Conservation District Advisory Committee: 1985, pp 18-23.

[26] Cransberg, L; McFarlane, DJ. Can perennial pastures provide the basis for a sustainable farming system in southern Australia? *New Zealand Journal of Agricultural Research*, 1994, 37, 287-294.

[27] Brack, C; Coops, N; Furby, S; James, R; Kesteven, J; Landsberg, J; Raison, J; Richards, G; Turner, B. *Report No. 27 - Biomass estimation: approaches for assessment of stocks and stock change*; Australian Greenhouse Office: *Canberra*, 2002.

[28] Cleugh, HA; Prinsley, R; Bird, RP; Brooks, SJ; Carberry, PS; Crawford, MC; Jackson, TT; Meinke, H; Mylius, SJ; Nuberg, IK; Sudmeyer, RA; Wright, AJ. The Australian National Windbreaks Program: overview and summary of results. *Australian Journal of Experimental Agriculture*, 2002, 42, 649-664.

[29] Ruysschaert, R; Poesen, J; Verstraeten, G; Govers, G. Soil loss due to harvesting of various crop types in contrasting agro-ecological environments. *Agriculture, Ecosystems and Environment*, 2007, 120(2-4), 153-165.

[30] Paustian, K; Andren, O; Janzen, HH; Lal, R; Smith, P; Tian, G; Tiessen, H; Van Noordwijk, M; Woomer, PL. Agricultural soils as a sink to mitigate CO2 emissions. *Soil Use and Management*, 1997, 13, 230-244.

[31] Sherrod, LA; Peterson, GA; Westfall, DG; Ahuja, LG. Cropping intensity enhances soil organic carbon and nitrogen in a no-till agroecosystem. *Soil Science Society of America*, 2003, 67, 1533-1543.

[32] Bronick, CJ; Lal, R. Soil structure and management: a review. *Geoderma*, 2005, 124, 3-22.

[33] Hatton, TJ; Nulsen, RA. Towards achieving functional ecosystem mimicry with respect to water recycling in southern Australian agriculture. *Agroforestry Systems*, 1999, 45, 203-214.

[34] Sanford, P; Wang, X; Greathead, KD; Gladman, JH; Speijers, J. Impact of Tasmanian blue gum belts and kikuyu-based pasture on sheep production and groundwater recharge in south-western Western Australia. *Australian Journal of Experimental Agriculture*, 2003, 43(8), 755-767.

[35] Patabendige, DM; Scott, PR; Lefroy, EC. *Fodder trees and shrubs for high rainfall areas of south Western Australia*. Department of Agriculture Western Australia: Perth, 1992.

[36] Resh, SC; Binkley, D; Parrotta, JA. Greater soil carbon sequestration under nitrogen-fixing trees compared with eucalyptus species. *Ecosystems*, 2002, 5(3), 217-231.

[37] Simpson, JR. Mineral nitrogen fluctuations in soils under improved pasture in southern New South Wales. *Australian Journal of Agricultural Research*, 1962, 13, 1059-1072.

[38] Weaver, DM; Prout, AL. Changing farm practice to meet environmental objectives of nutrient loss to Oyster Harbour. *Fertilizer Research*, 1993, 36, 117-184.

[39] Wise, R; Cacho, O. A bioeconomic analysis of soil carbon sequestration in agroforests. In Graduate School of Agricultural and Resource Economics (GSARE): 1999.

[40] Valzano, F; Murphy, B; Koen, T. *Report No. 43 - The impact of tillage on changes in soil carbon density with special emphasis on Australian conditions*; Australian Greenhouse Office: Canberra, 2005.

[41] Andre, HM; Ducarme, X; Lebrun, P. Soil biodiversity: myth, reality or conning? *Oikos*, 2002, 96(1), 3-24.

[42] Gustafsson, O; Haghseta, F; Chan, C; Macfarlane, J; Gschwend, PM. Quantification of the dilute sedimentary soot phase: implications for PAH speciation and bioavailability. *Environmental Science & Technology*, 1997, 31, 203-209.

[43] Accardi-Dey, A; Gschwend, PM. Assessing the combined roles of natural organic matter and black carbon as sorbents in sediments. *Environmental Science & Technology*, 2002, 36, 21-29.

[44] Lehmann, J; Pereira da Silva Jr, J; Steiner, C; Nehls, T; Zech, W; Glaser, B. Nutrient availability and leaching in an archaeological Anthrosol and a Ferralsol of the Central Amazon basin: fertiliser, manure and charcoal amendments. *Plant and Soil*, 2003, 249, 343-357.

[45] Mizuta, K; Matsumoto, T; Hatate, Y; Nishihara, K; Nakanishi, T. Removal of nitrate-nitrogen from drinking water using bamboo powder charcoal. *Bioresource Technology*, 2004, 95, 255-257.

[46] Giller, KE; Beare, MH; Lavelle, P; Izac, AMN; Swift, MJ. Agricultural intensification, soil biodiversity and agroecosystem function. *Applied Soil Ecology*, 1997, 6(1), 3-16.

[47] Tobor-Kaplon, MA; Bloem, J; De Ruiter, PC. Functional stablity of microbial communities from long-term stressed soils to additional disturbance *Environmental Toxicology and Chemistry*, 2006, 25(8), 1993-1999.

[48] Yan, F; McBratney, AB; Copeland, L. Functional substrate biodiversity of cultivated and uncultivated A horizons of vertisols in NW New South Wales. *Geoderma*, 2000, 96(4), 321-343.

[49] Mertens, J; Ruyters, S; Springael, D; Smolders, E. Resistance and resilience of zinc tolerant nitrifying communities is unaffected in long term zinc contaminated soils. *Soil Biology and Biochemistry*, 2007, 39(7), 1828-1831.

[50] Heemsbergen, DA; Berg, MP; Loreau, M; van Hal, JR; Faber, JH; Verhoef, HA. Biodiversity effects on soil process explained by interspecific functional dissimilarity. *Science*, 2004, 306(5698), 1019-1020.

[51] Auge, RM. Arbuscular mycorrhizae and soil/plant water relations. *Canadian Journal of Soil Science*, 2004, 84, 373-381.

[52] Wright, AL; Dou, F; Hons, FM. Soil organic C and N distribution for wheat cropping systems after 20 years of conservation tillage in central Texas. *Agriculture, Ecosystems and Environment*, 2007, 121, 376-382.

[53] Todd, JJ. Woody-biomass fuelled power stations. In *Renewable Energy Generators Association (REGA) Forum*, 2001, Launceston, 2001.

[54] Post, WM; Kwon, KC. Soil carbon sequestration and land-use change: processes and potential. *Global Change Biology*, 2000, 6(3), 317-327.

[55] Milne, E; Powlson, DS; Cerri, CE. Soil carbon stocks at regional scales (preface). *Agriculture, Ecosystems and Environment*, 2007, 122(1), 1-2.

In: No-Till Farming
Editor: Earl T. Nardali

ISBN: 978-1-60741-402-5

Chapter 6

CARBON SEQUESTRATION IN NO-TILLAGE SYSTEMS ON TROPICAL SOILS OF BRAZIL

Luiz Fernando Carvalho Leite[1*], Edvaldo Sagrilo[1], Sandra Regina Silva Galvão[1] and Giovana Alcântara Maciel[1]

[1]Embrapa Mid-North. Av. Duque de Caxias, 64006-220, Teresina-PI, Brazil.

ABSTRACT

The soil contains a significant part of the global carbon (C) stock: 3.5 % compared with to 1.7 % in the atmosphere, 8.9 % in fossil fuels, 1.0 % in biota and 84.9 % in oceans. With increasing emission of greenhouse gases (GHG) there is a growing interest in assessing the soil role as a C sink in different management conditions. The replacement of natural to agricultural ecosystems, with annual crops and based on plow tillage has resulted in decline in soil organic carbon (SOC) stocks. In tropical regions, where high temperatures and rainfall can increase the soil degradation, the use of no tillage (NT) can reverse this process, favoring the recovery of SOC, decreasing CO_2 emissions, and hence reducing the greenhouse effect. The widespread use of the NT in Brazil is characterized by sowing on to the crop residues from the previous cultivation, without any soil preparation. The dead biomass formed by the crop residues accumulated on the topsoil can help reduce erosion, temperature and moisture fluctuations at the surface, and the crops exposure to drought. These characteristics result in chemical, physical and biological alterations in soil properties affecting nutrient uptake by crops. In Brazil, the soils are dominated by variable-charge clay minerals such as kaolinite and, especially, Fe and Al oxides, which have a high flocculation capacity due to electrostatic interactions through their positive charges. The strong adsorption of SOC to active Al and Fe oxides is the cause for SOC stability in tropical soils; therefore, the interaction of organic compounds with clay in highly weathered oxidic soils can prevent SOC losses. In this context, soil C sequestration is a significant option to mitigate GHG emission in Brazilian tropical conditions

* Corresponding author (luizf@cpamn.embrapa.br).

INTRODUCTION

The terrestrial biosphere contains about 1500 to 1600 Pg of C in the top 1 m of soil and an additional 470 to 655 Pg of C in the vegetation (Lal, 2004). Together, C stocks in the terrestrial ecosystem are three times the amount of C in the atmosphere. The soil C stock in the top 30 cm is about 800 Pg C, which is almost the same amount as that stored in the atmosphere (Cerri et al., 2006).

The replacement of natural to agricultural ecosystems, with annual crops and based on plow tillage degrades soil quality by mixing surface soil with subsoil and leads to depletion of the soil organic carbon (SOC) stock. Mechanical disturbance by plow tillage disrupts aggregation, exposes SOC protected within the aggregates to microbial attack, and exacerbates erosion-induced soil degradation and increase the emission of the CO_2 to the atmosphere (Sá & Lal, 2008). In tropical regions, where high temperatures and rainfall can increase soil degradation process, the use of conservation management systems can reverse this process, favoring the recovery of SOC, decreasing CO_2 emissions, and hence reducing the greenhouse effect.

No tillage (NT) system is one of the most efficient practices for sequestering C in cropland and its effects depends on soil texture, being directly proportional to clay content and varies with soil zone and cropping frequency (Campbell et al., 2005). In Brazil, NT started in the south region (Paraná State) in 1972 as an alternative to the misuse of land causing erosion. The underlying land management principles that led to the development of NT system in Brazil were, prevention surface sealing caused by rainfall impact, achievement and maintenance of an open soil structure and reduction of the volume and velocity of surface runoff. Consequently, the no-tillage strategy was based on two essential farm practices: (i) not tilling and (ii) keeping soil covered at all times (Cerri et al., 2007).

Higher C stocks in NT than in CT have been reported by several Brazilian studies (Leite et al., 2003; Amado et al., 2006; Bayer et al; 2006), but this trend was not observed by other ones (Leite et al., 2009). These contrasting results are related to the diverse cropping system associated with no-tillage. NT with a second crop has more potential to sequester C than no-tillage systems with only one crop per year. Moreover, soil texture and mineralogy also play an essential role. Clayey soils have a larger proportion of stabilized soil organic matter (SOM) which will be less affected upon tillage compared to sandy soils. In Brazil, the soils are dominated by variable-charge clay minerals such as kaolinite and, especially, Fe and Al oxides, which have a high flocculation capacity due to electrostatic interactions through their positive charges. Numerous authors have emphasized that the strong adsorption of SOC to active Al and Fe oxides is the cause for SOC stability in tropical soils; therefore, the interaction of organic compounds with clay in highly weathered oxidic soils can prevent SOC losses. In this context, soil C sequestration is a significant option to mitigate GHG emission in Brazilian tropical conditions

Soil Management Systems and Carbon Sequestration Rate

Soil is an important terrestrial C reservoir which plays a significant role in the global C cycle and can function either as a source or a sink for atmospheric GHG depending on land use and soil management (Zanatta et al., 2007; Ussiri and Lal, 2009). According to Hutchinson et al. (2007), 20-30% (temperate region) and 50-75 % (tropical region) of native C stock can be released to the atmosphere within the first 20 years of cultivation.

Table 1. Amount of C sequestered in agricultural soils for various management practices (Adapted from Hutchinson et al., 2007)

Practice	Amount of C sequestered (range) (Mg C ha^{-1} $year^{-1}$)	Region
Nutrient addition via fertilizer	0.30	Forty-two data points
	0.05-0.15	USA
	0.14-0.18	USA
	0.16	USA
	0.04	Canada
Nutrient additions via manure	0.2-0.5	USA
Converting cultivated land to grasslands	1.01	Twenty-three data points
	0.62	Canada
Including more forages (especially legumes) in rotation	0.75	Six data points
	0.44	Canada
Adopting conservation tillage	0.06	Canada
	0.3-0.6	USA
	0.1-0.5	USA
Conversion from conventional to no-till	0.13	Canada
	0.20-0.35	Canada
	0.14-0.56	USA
	0.5-0.8	USA
	0.17-0.56	Div. countries
	0.12-0.29	Asia
	0.09-0.29	Africa
	0.12-0.29	L. America
Reduction of summer fallow	0.123-0.185	Canada
	0.07	Canada
	0.03-0.14	Canada
	0.02-0.04	Canada
	0.05-0.4	USA
Improved grassland management	0.5-0.6	Australia
	0.59	Canada/US
	0.28	Australia
	0.05-0.3	USA

Different agricultural practices are known for their capacity to increase C sequestration and to reduce GHG emission (Hutchinson et al., 2007). Minimum tillage, sustainable cropping systems (crop rotation, intercropping, agroforestry), balanced organic and inorganic fertilizers and grasslands improvement are recognized as effective practices to build up and maintain higher levels of C in soils. More recently, other potential alternatives have been highlighted, as the use of pyrolized biomass (biochar) applied directly to the soil, resulting in stable C sequestered in soil and improving its agronomic quality.

Extensive research has shown clearly that is possible to combine for each situation, management practices capable of increase SOC stocks and mitigate CO_2 emissions. Such practices are usually followed by improvement in soil quality and maintain high and sustainable levels of crop production (Table 1).

Cropping Systems

Crop rotation based on legumes and cover crops in the rotation as well as nutrient management and conservation tillage are effective to preserve soil fertility and crop productivity through improvement in soil C sequestration (Nieder and Bendi, 2008). At northwest China, Li et al. (2009) reported that, in a 10-year period, SOC stock increased 30-65% in the 0-30 soil depth resulting in a C sequestration rate of 0.6, 0.3 and 0.8 Mg C ha^{-1} $year^{-1}$ under wheat-maize, wheat-alfalfa and wheat-acacia rotations, respectively.

In crop sequence which there is no legume species in the rotational scheme has been also observed an increase in the SOC stock. Kukal et al. (2009) observed that SOC stock in rice–wheat sequence was 54 and 30% higher than in the maize-wheat sequence, with the application of farmyard-manure and balanced NPK fertilization. This result can be attributed to greater aboveground biomass production and favorable water regime during rice season. However, the inclusion of legumes in the cropping rotations returns large amounts of crop residues to the soil and can potentially increase SOC stock. Moreover, the biological nitrogen fixation enhanced by legumes provides high N-content plant residues, reducing requirements for inorganic-N, thereby, reducing the C cost of manufacturing N fertilizers (Hutchinson et al., 2007).

Zanatta et al. (2007) in Southern Brazil, reported that the legume-based cropping systems increased the C addition to the soil by 40% in the vetch-maize (V/M) sequence and 87% in the oat+vetch-maize+cowpea (OV/MC) sequence compared to the grass-based sequence (oat-maize) in treatments without inorganic-N fertilization and by 9% (V/M) and 37% (OV/MC) in the treatments that received N fertilization. The same trend was observed for SOC stocks with higher values for OV/MC (5.8 Mg ha^{-1}) and V/M (5.1 Mg ha^{-1}), compared to O/M sequence, in the 0–30 cm soil layer. These results emphasize the benefits of legume-based and multiple cover crop-based systems to increase SOC stocks in subtropical regions.

Southern Brazil has a well-distributed rainfall pattern through the year allowing for the development of intensive cropping system with three or even for crops per year, which can result in high addition of crop residue to the soil surface. These systems increase the input of biomass C to the soil and, consequently lead to the accumulation of C in soils under conservation tillage systems (Amado et al., 2006; Bayer et al., 2006). Other Brazilian regions which include Cerrados (Brazilian Savannah) and Amazon basin, show a seasonal rainfall pattern with a dry season, what reduces options for intensive rotational cropping systems. At those regions, the conversion of forests or cropland into perennial grassland for forage production can result in a substantial increase in soil C sequestration (Hutchinson et al., 2007). Moraes et al. (2002) compared soil C stock of two chronosequences of 8 and 20 years old pastures and native forest in the west region of Amazon basin, Brazil. In one of the two chonosequences they observed higher SOC stocks in the 30 cm depth in pastures than in the original forest. There was no difference among land management in the other chronosequence. SOC stock derived from C4 pasture grasses increased with pasture age, but a large portion of the SOC stock in both chronosequences consisted of residual forest C. Carbon derived from the original forest comprised 64% to 83% of total soil C at the 0 to 10 cm depth in 8-year-old pastures and 48% to 50% in 20-year-old pastures.

Currently, pasture area in Brazil is estimated to reach 172.3 millions hectares (IBGE, 2006). Maia et al. (2009) suggest that about 50 to 80% of the pastures, mainly at the cerrados and Amazon basin, are considered degraded and therefore, the adoption of nominal or improved management practices would reverse the C losses trends. However, C sequestration data do not consider the high amounts of C-emission from slash-and-burn process, typically used for the conversion of forest to grassland or croplands. Besides, it is widely known in Brazil the idea that it is not necessary to open new areas of forest to increase cattle production.

Integrated Crop-Livestock System (ICLS) is an available technology that has been adopted mainly for farmers from the Cerrado region. Grasses (often Brachiaria sp.) are sown in associated with the main commercial crop to achieve a dense soil cover at harvest of the main crop. The resulting pasture is an excellent cover crop and can be maintained during a few years. Before pasture degradation, the field is returned to cropland by direct sowing of a commercial crop into the mulch of pasture residues after a total-herbicide application (Bustamante et al., 2006). There are four positive factors associated with livestock being integrated into cropping enterprises: (i) crops produced on the farm can be used to feed the livestock, thus minimizing the importing of outside feed stuffs in livestock production; (ii) livestock manure can serve as the primary source of nutrients for crop production, thereby cycling nutrients from the crops through the animals and back out onto the land; (iii) livestock can serve as the sink for agricultural byproducts; and (iv) ruminant livestock encourage the establishment of perennial grass and legume forages as a primary feedstuff (Randal, 2003). In spite of its importance, little is known about the effect of the ICLS on the SOC accumulation and C sequestration.

Soil Amendments

Several authors have reported an increase in SOC stocks as a function of inorganic fertilizers application (Purakayastha et al, 2008; Sainju et al., 2008; Gong et al., 2009; Kukal et al., 2009) and a reduction on CO_2 emission to atmosphere (Iqbal et al., 2009). Hutchinson et al. (2007) reported that many agricultural ecosystems are N-limited and that adding N fertilizer usually results in increased crop production and C sequestration in soils as a consequence of C inputs by improved biomass production. However the authors emphasized that for each 1 kg of manufactured and transported N, about 1 kg C is released to the atmosphere. For the account of the net effect of N fertilization on C sequestration, this information must be considered, as also reported by Purakayastha et al. (2008). At South

Brazilian conditions, Zanatta et al. (2007) observed positive effect of added inorganic N to soil on SOC stocks. However, when C-costs related to N-based fertilizers were taken into account, the increase in SOC accumulation due to N did not necessarily improve atmospheric C mitigation. On the other hand, application of organic amendments to soil seems to be a more sustainable practice.

Purakayastha et al. (2008) verified greater increases in SOC stock and C sequestration (0. 97 Mg C ha^{-1} $year^{-1}$) when farmyard manure was applied in combination with smaller rates of NPK, compared to sole NPK at higher rates. Sodhi et al. (2009) observed that the rice straw applied alone or in combination with inorganic fertilizers increased soil macro-aggregate and C sequestration. Similar results were reported by Gong et al. (2009) and Kukal et al (2009).

In Brazil, Gomes et al. (2005) found linear response of SOC stock to maize+cattle manure compost application in a two-year corn grain production system. An increase of 0.0076% in the SOC was obtained for each cubic meter of added compost. This result was expected since frequent organic fertilization represents a considerably C input to the soil. Such statement is corroborated by Araújo et al. (2008) data, in a perennial organic production system of *Malpighia glaba* at northeast Brazil. The authors verified a build-up of 53% in the SOC stocks in a two-year period in a Typic Quartzipsamment witch was attributed to the high inputs of organic matter.

More recently, other efforts to improve soil quality and simultaneously increase SOC stock have been carried on. In this context Biochar (black carbon) application to the soil has emerged as a novel approach to establish a significant long-term sink for atmospheric carbon dioxide in terrestrial ecosystems (Lehmann et al., 2006). Biochar results from pyrolizated biomass. It is relatively recalcitrant and has a long residence time in soil, a possible option to use it as a nutrient-fortified soil Amendment (Lal, 2008). The concept of land application of biochar is not recent and derives from the pre-Colombian anthropogenic soils, so-called Amazonian Dark Earths (*Terra Preta de Índio*), where the total C storage is as high as 250Mg C ha^{-1} m^{-1}. Biochar can act as a soil conditioner enhancing plant growth by supplying and, more importantly, retaining nutrients and improving soil physical and biological properties, as reported by Steiner. (2007). Available data suggests that crops respond positively to bio-char additions up to 50 Mg C ha^{-1} and may show growth reductions only at very high applications. For most plant species and soil conditions, this maximum was not reached even with 140 Mg C ha^{-1} (Lehmann et al., 2006), witch represents a strong and long-term stable sink for C in tropical soils.

Soil Tillage

No till (NT) management is one of the most efficient practices for sequestering C in cropland (Hutchinson et al., 2007). Recent research has shown potential effects of no-till or reduced tillage on SOC sequestration in several countries (Yang et al., 2008 in Canada and USA; Spargo et al, 2008 in USA; Ussiri and Lal, 2009 in USA; Jacobs et al., 2009 in Germany; Melero et al., 2009 in Spain; Chatterjee and Lal, 2009 in USA)

Globally, approximately 63 million hectares of land are presently under NT systems with the USA having the largest area at about 21 million hectares (Derpsch, 2001). In Brazil, NT systems started in the south region in 1972 as an alternative to the misuse of land, which had caused erosion. This alternative quickly expanded to different states and the area under no tillage has since then increased exponentially (Cerri et al.,2004). Nowadays, about 18 millions ha covered by no till make Brazil the second largest adopter in the world.

The widespread use of the NT in Brazil is characterized by sowing on to the crop residues from the previous cultivation, without any soil preparation. The dead biomass formed by the crop residues accumulated on the topsoil can help reduce erosion, temperature and moisture fluctuations at the surface and the crops exposure to drought. These characteristics result in chemical, biological and physical alterations in soil properties.

NT improves physical protection of SOM within soil aggregates. Particulate fractions of SOM occluded into soil aggregates have longer turnover times than free fractions. However, few studies have been performed in soils from southern Brazil. Variable charge minerals could improve this mechanism of SOM stabilization and as a consequence contribute to C accumulation in NT soils. In general, the less oxidative environment in NT soils increase the lability of SOM as evidenced by spectroscopic techniques and by the ratio between labile and non-labile pools of SOM (Amado et al., 2006)

No-Tillage Can Increase Carbon Sequestration in Tropical Soils

Intensive and continued tillage practices have caused enormous losses of SOC and N pools as greenhouse gases to the atmosphere. It is estimated that as much as 60% of SOC in temperate regions and 75% in the tropics has been depleted by conventional tillage (CT), contributing about 23% of the total greenhouse gas concentration in the atmosphere (IPCC, 1966, Lal, 2004; Blanco-Canqui et al., 2008).

There are several mechanisms which can promote soil carbon loss in CT: disruption of soil aggregates, which protect soil organic matter from decomposition; stimulation of short-term microbial activity through enhanced aeration, resulting in increased levels of CO_2 and other gases released to the atmosphere and; mixing of fresh residues into the soil where condition for decomposition are often more favorable than on the surface (Cerri et al., 2004)

As a consequence, NT farming is being promoted as an alternative to CT for restoring SOC as an ancillary benefit of NT farming. It is generally viewed that switching from CT to NT would restore the SOC pool that has been lost thereby offsetting emissions by fossil fuel combustion and alleviating concerns of the projected global climate change (West and Post , 2002; Blanco-Canqui et al., 2008). Increase in the SOC stocks, in temperate soils, with NT compared CT have often been observed concomitant with improved soil aggregation According to Six et al. (2000), increases in total SOC under NT relative to CT management are attributed not only to a greater amount of C-rich macroaggregates (>250 μm), but also to a reduced rate of macroaggregate turnover under NT. The slower turnover of macroaggregates was suggested to enhance the formation of highly stable microaggregates (within macroaggregates) in which C is stabilized and sequestered in the long term (Denef et al., 2007). Barthés et al (2008) emphasized that the stability of SOC and aggregation is affected by the presence of iron and aluminum oxides which cover about 60-70 % of tropical soils.

It has been argued that widespread adoption of NT within US could sequester 24-40 Mt C $year^{-1}$(Lal, 2003). These statistics have been projected globally to estimate that conversion of all cropland to NT could sequester 25 Gt C over the next 50 years, marking it as one of the key global strategies for stabilizing atmospheric CO_2 concentration (Baker et al., 2007). The time needed for a NT system to sink CO_2 into soils is variable, depending on many factors. It can be generally assumed that most of the C sequester due to changes in the management systems will continue for 25 - 50 years (Follett 2001), highlighting that 15 years appears to be not long enough for reaching an equilibrium between the rate of C sequestration and C-loss.

In the tropics, specifically in Brazil, the rate of carbon accumulation has been estimated in the two main regions (south and centre-west) under NT system. In the southern, Sá (2001) and Sá et al (2001) estimated higher sequestration rates of 0.8 Mg C ha^{-1} yr^{-1} in the 0-20 cm layer and 1.0 Mg C ha^{-1} yr^{-1} at 0-40 cm depth after 22 years under NT compared with soils under CT over same period. The accumulated C was generally greater in the coarse (> 20μm) than in the fine (<20 μm) particle-size fraction indicating that most of this additional C is weakly stable. Amado et al (2006), in the same region, studied several soil management and crop systems and reported that NT show a range of 0.12 to 0.43 Mg C ha^{-1} yr^{-1} of C accumulation relative to the CT soil. In the sand clay loam Paleudult, NT had 0.12 Mg ha^{-1} yr^{-1} C higher accumulation than CT under the double-cropping system (black oat-maize). In this soil, SOC increment rates were 26.7 (0-5 cm) and 6.0 (0-20 cm) kg ha^{-1} cm^{-1} yr^{-1}, indicating that C accumulation in NT soils occurred preferentially in soil surface layer. In the clay Hapludox, NT had 0.16 Mg ha-1 yr-1 of C accumulation compared to CT, with double-cropping system (wheat-soybean). TOC increment rates were 39 (0-5cm) and 8 (0-20 cm) kg ha^{-1} cm^{-1} yr^{-1}. According to the authors, the similar rate of NT C accumulation in the clay Hapludox (0.16 Mg C ha^{-1} yr^{-1}) in comparison to sandy clay loam Paleudult (0.12 Mg C ha^{-1} yr^{-1}) probably was related to higher C addition in the sandy clay loam Paleudult than in clay Hapludox which partly compensates for the lower C stabilization in the Paleudult (Table 2)

In the Central Region of Brazil, especially Cerrado biome, several authors have reported an increase in the SOC stocks under NT system (Leite et al., 2003; Bayer et al., 2004; Bayer et al., 2006) but this was not observed by other ones (Roscoe and Buurman, 2003). These contrasting results are probably related to the diverse cropping practices associated with no-tillage. It is clear that NT systems with a second crop have more potential to sequester C than NT systems with only one crop per year. In addition, differences in soil texture also play a role. Clayey soils have a larger proportion of stabilized SOC which will be less affected upon tillage compared to sandy soils (Bustamante et al., 2006).

Bayer et al (2006) studied the role of two representative Cerrado Oxisols (350 and 650 g clay kg^{-1}) as source or sink of C atmospheric C managed under conventional, reduced and no-tillage. Higher SOC stocks were verified in NT compared to tilled soils (which both had similar C input from the same cropping system) and attributed to a lower rate of organic matter decomposition when soils are left undisturbed. These results highlight the importance of physical protection of organic matter within stable aggregates of NT soils in increasing C stocks and C sequestration. Besides, the authors reported that changing the soil management scheme from CT to NT resulted in an annual C sequestration rate of 0.3 Mg ha^{-1} $year^{-1}$ in the sandy clay loam Oxisol and of 0.60 ha^{-1} $year^{-1}$ in the clayey Oxisol and calculated an average C sequestration rate of 0.35 Mg C ha^{-1} $year^{-1}$ for the 0-20 cm layer of Brazilian tropical no-till soils (Table 3)

Table 2. Effects of soil management and crop system on annual rate of carbon accumulation in four different soil types from southern Brazil. (Adapted from Amado et al., 2006)

Soil	Clay content	Time of adoption	Crop system	SOC * (0-20 cm)		Carbon accumulation		
				CT*	NT*	Tillage effect	Cropping effect	Tillage and cropping effect
	g Kg^{-1}	yr		-Mg ha^{-1}-		Mg ha^{-1} yr^{-1}		
Sandy loam Paleudalf	87	10	mono-cropping (fallow-maize)	-	19.2	-	-	-
			double-cropping (ryegrass-maize)	-	20.8	-	0.43 (TLCC x double-cropping under NT)	
			TLCC** (velvet beans-maize)	-	25.1	-	0.59(TLCC x mono-cropping under NT)	-
Sandy clay loam Paleudult	220	15	double-cropping (black oat-maize)	33.8	35.6	0.12 (NT x CT under double-cropping)	-	-
			ICS I§	-	39.4	-	0.25 (intensive x double-cropping under NT)	0.37 (NT intensive x CT double-cropping)
			TLCC (pigeon pea-maize)	41.3			0.38 (TLCC x double-cropping under NT)	
Clay Hapludox	570	19	double-cropping (wheat-soybean)	48.7	51.8	0.16 (NT x CT under double-cropping)	0.26 (intensive x double-cropping under CT)	-
			ICS II§	53.6	58.3	0.25 (NT x CT under ICS II)	0.34 (intensive x double-cropping under NT)	0.51 (NT intensive x CT double-cropping)
High clay Hapludox	760	7	ICSIII§	50.4	53.4	0.43 (NT x CT under ICS III)	-	-

*TOC, total organic carbon; CT, conventional tillage; NT, no-tillage.

** TLCC, tropical legume cover crop.

§Intensive Cropping System (ICS) I = black oat + common vetch-maize + cowpea. ICSII = black oat-soybean-black oat + vetch-maize-radish oil-wheat-soybean. ICS III = triticale-rye-soybean-common vetch-maize-black oat-black bean-buck wheat-radish oil.

Table 3. Carbon sequestration rates in Brazilian tropical and subtropical no-till soils (adapted from Bayer et al (2006).

Site	Soil	Climate[a]	Cropping System[b]	Depth (cm)	C stock ($Mg\ ha^{-1}$)		Time (year)	Rate ($Mg\ ha^{-1}\ year^{-1}$)	Reference
					CT	NT			
Tropical				0-20				0.35	
Embrapa CPAC Planaltina, DF	Oxisol 41% clay	- -	Win: F Sum: R, S, P, F and M	0-20	37.3 37.0	47.4 47.5	16 20	0.63 0.52, 0.58	Corazza et al. (1999) Oliveira et al. (2004)
Agropec. SLC Luziânia, GO	Oxisol 35% clay	23.8 1581	Win: F Sum: S and M	0-20	38.6	41.0	8	0.30	Bayer et al. (2006)
Emater-GO Sen. Canedo, GO	Oxisol 50% clay	23.0 1500	Win: F Sum: M and B	0-20	48.6	49.6	4	0.25	Freitas et al. (2000)
Agropec. SLC Costa Rica, MS	Oxisol 65% clay	24.8 1929	Win: F Sum: S and M	0-20	54.3	57.3	5	0.60	Bayer et al. (2006)
Embrapa CNPMS Sete Lagoas, MG	Oxisol 80% clay	22.1 1340	Win: F Sum: M and B	0-20	50.3	50.0	10	-0.03	Roscoe and Buurman (2003)
UFV Coimbra, MG	Acrisol 40% clay	19.5 1350	Win: F Sum: M/F/S	0-20	31.2	38.5	15	0.49	Leite et al. (2004)
IAC Campinas, SP	Oxisol 60% clay	23.0 1060	O/S; O/M-O/S; O/S-O/M; O/M (win/sum)	0-20	78.0	80.8	12	0.23	DeMaria et al. (1999)

[a] The values refer to the mean annual temperature (8C) and mean annual precipitation (mm), respectively.

[b] win (winter), sum (summer), B (beans, *Phaseulus vulgaris*), F (fallow), Lu (lupin, *Lupinus* sp.), M (maize, *Zea mays*), O (oat, *Avena strigosa*), P (pigeon pea, *Cajanus cajan*), R (rice, *Oriza sativa*), S (soybean, *Glycine max*), V (vetch, *Vicia sativa*)

[c]: C stocks (per volume) are not shown.

Table 4. Soil C stocks (Mg ha^{-1}) and C sequestration rate (Mg ha^{-1} year^{-1}) in the equivalent layers of 0 – 10 and 0 – 30 cm in the dry and wet seasons under Cerrado (CE), conventional tillage (1CT and 2CT) and no-tillage (1NT, 2NT and 3NT) in Vilhena, Rondonia State, Brazil. (Adapted from Carvalho et al.2009).

Area	Soil C Stocks (Mg ha^{-1})			
	Soil equivalent layer (10 cm)		Soil equivalent layer (30 cm)	
	Dry season	Wet season	Dry season	Wet season
CE	21.5 a	21.3 a	50.0 b	48.1 a
1CT	16.3 b	17.8 a	47.6 b	47.4 a
2CT	21.3 ab	24.2 a	55.4 ab	58.5 a
1NT	23.3 a	24.2 a	66.5 a	65.6 a
2NT	21.3 ab	19.0 a	54.5 ab	47.4 a
3NT	23.9 a	19.5 a	67.5 a	59.0 a
LSD[a]	5.14	10.6	14.5	22.9
CV%[b]	12.4	25.7	13.0	21.5

The results are mean (n=5) ± standard deviation. Means followed by the same letter are not significantly different according to Tukey's test at 5%.

[a] Least significant difference.

[b] Coefficient of variation.

More recently, Carvalho et al (2009) showed that during dry season significant differences in soil C stocks as a function of land use change were observed (Table 4). After converting Cerrado into agricultural land, SOC stock in CT cultivated with rice for 1 year (1CT) was 16.3 Mg ha^{-1} smaller ($p<0.05$) than the stocks in Cerrado (CE), no-till cultivated with soybean for 1 year (1NT) and no-till cultivated with soybean for 2 years (2NT). NT systems were installed after 2 years of rice under CT. SOC stocks in the 0-10 cm layer under NT were 23.3, 21.3 and 23.9 Mg ha^{-1}, for the 1NT, 2NT and 3NT, respectively (Table 3). Higher C stocks in the NT compared to CT were probably due to the absence of tillage operations (i.e., NT). The lower decomposition rates favored the maintenance and accumulation of SOM and guaranteed a continuous flow of subtracts and energy through the soil (Green et al., 2007; Carvalho et al., 2009).

Leite et al (2009) estimated C sequestration in two Cerrado soils (Ultisol and Oxisol) under no-tillage and plowed systems using CQESTR model. For the Ultisol, CQESTR simulated that tillage systems are emitting 0.36 and 1.05 Mg C ha^{-1} year^{-1} under NT and CT, respectively. Measured values also indicated that all tillage systems (including NT) emitted C; however, they were smaller in magnitude (Table 5). These estimates differ from others reported elsewhere showing carbon sequestration under no-tillage systems. In this study it was used C stocks at the beginning of the experiment as a reference for calculations. On the other hand, if estimates for 2000 were based on C stocks for NT and CT, C sequestration on the order of 0.47 (simulated) and 0.48 Mg ha^{-1} year^{-1} (measured) would be reported for the 20-cm depth. When estimates are based on the difference between NT and CT, they agree well with published sequestration rates of 0.52 Mg ha^{-1} year^{-1} for Ultisols in Brazil (Lovato et al., 2004).

In the Oxisol, C emissions were also observed, and varied from 0.30 (NT) to 0.82 (CT) when estimated by CQESTR, and from 0.14 (NT) to 0.84 Mg ha^{-1} year^{-1} (CT) when calculated from measured values. Like the Ultisol, the highest difference between simulated and measured values was observed in the no-tillage system (53%) compared with conventional (2%) and reduced (4%) tillage systems (Table 5). Considering SOC stocks in conventional tillage plots as the baseline, in 2006, NT estimated and measured annual C sequestration rates were 0.18 and 0.38 Mg ha^{-1} year^{-1}, respectively. In fact, C sequestration rates in Brazilian Oxisols vary depending on clay content and number of years of the experiment. Freitas et al. (2000) estimated an annual C sequestration rate of 0.25 Mg ha^{-1} year^{-1}(50% clay; 4 years) while Bayer et al. (2006) calculated 0.30 Mg ha^{-1} year^{-1}(35% clay; 8 years), which is similar to 0.35 Mg ha^{-1} year^{-1} (50% clay; 5 years) observed by Metay et al. (2006). In longer experiments, Corazza et al. (1999) and Oliveira et al. (2004) observed a C sequestration rate of 0.50 (40–50% clay; 15 years) and 0.52 Mg ha^{-1} year^{-1} (41% clay; 20 years), respectively.

Table 5. Carbon emission, decomposition rate (k), and Mean Residence Time (MRT) estimated by CQESTR and by measured values in an Ultisol (Coimbra, Minas Gerais state) under no tillage (NT), reduced tillage (RT1: disk plow; RT2: heavy disk harrow), and conventional tillage (CT) and in an Oxisol (Baixa Grande do Ribeiro -BGR, Piauí state) under NT, reduced tillage (RT), and CT at 0–20 cm depth.

Tillage system	SOC stock			Rate	C Emission	k	MRT
	Initial	Final	Variation Δ				
		Mg ha^{-1}		Mg ha^{-1} year^{-1}			Year
	Ultisol (Coimbra)						
CQESTR							
NT	43.4	34.79	-8.60	-0.53	-0.36	0.014	72
RT1	43.4	28.45	-14.94	-0.93	-0.96	0.026	37
CT	43.4	27.18	-16.21	-1.01	-1.05	0.029	35
RT2	43.4	28.80	-14.59	-0.91	-0.94	0.026	39
Measured							
NT	44.0	38.54	-5.46	-0.34	-0.17	0.008	128
RT1	44.0	31.23	-12.77	-0.79	-0.82	0,021	49
CT	44.0	30.90	-13.10	-0.82	-0.86	0.022	48
RT2	44.0	31.24	-12.76	-0.79	-0.82	0.021	50
	Oxisol (BGR)						
CQESTR							
NT	42.7	36.07	-6.63	-0.51	-0.30	0.013	77
CT	42.7	32.53	-10.17	-0.78	-0.82	0.021	48
RT	42.7	33.75	-8.95	-0.69	-0.72	0.018	55
Measured							
NT	45.1	40.68	-4.42	-0.34	-0.14	0.009	126
CT	45.1	34.72	-10.38	-0.80	-0.84	0.022	49
RT	45.1	35.76	-9.34	-0.72	-0.75	0.019	56

(1) Contribution of 6 Mg ha^{-1} from cover crop (0.17 and 0.20 Mg ha^{-1} year^{-1} for Ultisol and Oxisol, respectively, considering 45% C) was assumed in no-tillage system (Salton et al., 2005) and a fuel additional C emission of 0.045 Mg ha^{-1} year^{-1} and 0.031 Mg ha^{-1} year^{-1} for conventional and reduced tillage, respectively (Dieckow, 2004).

While NT farming is highly beneficial to soil and water conservation and reduction of production costs, its potential for sequestering SOC needs a critical and objective assessment

(Puget and Lal, 2005). Baker et al. (2007) argued that the higher SOC sequestration in NT systems reported in many studies may merely be due to sampling protocol that could have biased the results. (Blanco-Canqui and Lal, 2008). The argument is that almost all the studies reporting higher SOC in NT soils relative to plowed systems have based their conclusions from samples collected within only a <30 cm soil depth and the few studies reporting SOC for the whole soil profile have found either no difference in SOC below the 30 cm depth or even lower SOC in NT than conventional systems (Baker et al., 2007; Blanco-Canqui and Lal, 2008). In most of the studies have been reported, in the superficial layer, higher SOC in NT system; however below to 30 cm and in the total profile, conventional system showed higher values (Lal, 1997). The lower SOC in deeper NT layers may thus offset the greater SOC in upper layers. Therefore, the total profile SOC between NT and CT soil management may not be significantly different.

Conclusion

Sustainable cropping systems with high biomass production, balanced organic and mineral fertilization and no tillage systems can increase SOC stocks and mitigated GHG emission in the tropical regions. Several studies have indicated higher SOC stocks under NT compared to CT. However, this trend occurs mainly in the surface soil layer (<10 cm) and no differences have been observed at the whole profile. In Brazil, an additional soil carbon accumulation due to NT of the 0.4-0.5 Mg ha^{-1} yr^{-1} for the 0-20 cm layer has been observed. NT minimizes SOC losses and is a promising strategy to maintain and increase SOC stocks but its role to carbon sequestration needs to study more carefully at different tropical regions.

Acknowledgments

The authors thank CNPq for the fellowship awarded to Luiz Fernando Carvalho Leite

References

Amado, T. J. C., Bayer, C., Conceição, P. C., Spagnollo, E., Campos, B. H. C. & Veiga, M. Potential of carbon accumulation in no-till soils with intensive use and cover crops in Southern Brazil. *Journal Environmental Quality*, *35*, 1599-1607.

Araújo, A. S. F., Santos, V. B. & Monteiro, R. T. R. (2008). Responses of soil microbial biomass and activity for practices of organic and conventional farming systems in Piauí state, Brazil. *European Journal of Soil Biology*, *44*, 225-230.

Baker, J. M., Ochsner, T. E., Venterea, R. T. & Griffis, T. J. (2007). Tillage and soil carbon sequestration: what do we really know? *Agriculture, Ecosystems and Environment, 118*, 1-5.

Barthes, B. G., Kouakoua, E. B., Larre-Larrouy A, M. C., Razafimbelo, T. M. C., De Luca, E., Azontonde, A. E., Neves, C. S. V. J., Freitas, P. L. & Felle, C. F. (2008). Texture and sesquioxide effects on water-stable aggregates and organic matter in some tropical soils Geoderma, 143, 14-25.

Bayer, C., Martin-Neto, L., Mielniczuk, J. & Pavinato, A. (2004). Armazenamento de carbono em frações lábeis da matéria orgânica de um Latossolo Vermelho sob plantio direto. *Pesquisa Agropecuária Brasileira*, *39*, 677-683.

Bayer, C., Martin-Neto, L., Mielniczuk, J., Pavinato, A. & Dieckow, J. (2006). Carbon sequestration in two Brazilian Cerrado soils under no-till. *Soil & Tillage Research, 86*, 237-245.

Blanco-Canqui, H. & Lal, R. (2008). No-tillage and soil-profile carbon sequestration: an on-farm assessment. *Soil Science Society of America Journal, 72*, 693-701.

Bustamante, M. M. C., Corbeels, M., Scopel, E. & Roscoe, R. (2006). Soil carbon storage and sequestration potential in the Cerrado Region of Brazil. In: LAL, R.

Campbell, C. A., Janzen, H. H., Paustian, K., Gregorich, E. G., Sherrod, L., Liang, B. C. & Zentner, R. P. (2005). Carbon storage in soils of the North American great plains: effect of cropping frequency. *Agronomy Journal, 97*, 349-363.

Carvalho, J. L. N., Cerri, C. E. P., Feigl, B. J., Píccolo, M. C., Godinho, V. P. & Cerri, C. C. (2009). Carbon sequestration in agricultural soils in the Cerrado region of the Brazilian Amazon. *Soil & Tillage Research*, *103*, 342-349.

Cerri, C. C., Bernoux, M., Cerri, C. E. P. & Feller, C. (2004). Carbon cycling and sequestration opportunities in South America: the case of Brazil. *Soil Use and Management*, *20*, 248-254.

Cerri, C. C., Bernoux, M., Cerri, C. E. P. & Lal, R. (2006). Challenges and opportunities of soil carbon sequestration in Latin America. In: LAL, R., CERRI, C. C., BERNOUX, M., Etchevers, J. & Cerri, C. E. P. (Eds). *Carbon Sequestration in Soils of Latin America*, pp. 41-47.

Cerri, C. C., Bernoux, M., Etchevers, J. & Cerri, C. E. P. (Eds). *Carbon Sequestration in Soils of Latin America*, pp. 285-304.

Cerri, C. E. P., Easter, M., Paustian, K., Killian, K., Coleman, K., Bernoux, M., Falloon, P., Powlson, D. S., Batjes, N., Milne, E. & Cerri, C. C. (2007). Simulating SOC changes in 11 land use change chronosequences from the Brazilian Amazon with RothC and Century models. Agriculture, *Ecosystems and Environment, 122*, 46-57.

Chatterjee, A. & LAL, R. (2009). On farm assessment of tillage impact on soil carbon and associated soil quality parameters. *Soil & Tillage Research, 104*, 270-277.

Corazza, E. J., Silva, J. E., Resck, D. V. S. & Gomes, A. C. (1999). Comportamento de diferentes sistemas de manejo como fonte ou depósito de CO em relação à vegetação de cerrado. *Revista Brasileira de Ciência do Solo*, *23*, 425-432.

Demaria, I. C., Nnabude, P. C. & Castro, O. M. (1999). Long-term tillage and crop rotation effects on soil chemical properties of a Rhodic Ferralsol in southern Brasil. *Soil & Tillage Research, 51*, 71-79.

Denef, K., Zotarellia, L. C., Boddey, R. & Six, D. J, (2007). Microaggregate-associated carbon as a diagnostic fraction for management-induced changes in soil organic carbon in two Oxisols *Soil Biology & Biochemistry*, 39, 1165-1172.

Derpsch, R. (2001). Frontiers in conservation tillage and advances in conservation practice. In: Stott, D. E., Mohtar, R. H. & Steinhardt, G. C. (Eds). Sustaining the Global Farm. Selected Papers from The 10th International Soil Conservation Organization Meeting held 24-29 May 1999. National Soil Erosion Research Laboratory, Purdue University and the

United States Department of Agriculture Agricultural Research Service (USDA-ARS), *USA*, p. 248-254.

Follet, R. F. (2001). Soil management concepts and carbon sequestration in cropland soils. *Soil & Tillage Research*, *61*, 77-92.

Freitas, P. L., Blancaneaux, P., Gavinelli, E., Larre′-Larrouy, M. C. & Feller, C. (2000). Nível e natureza do estoque orgânico de latossolos sob diferentes sistemas de uso e manejo. *Pesquisa Agropecuária Brasileira*, *35*, 157-170.

Gomes, J. A., Scapim, C. A., Braccini, A. L., Vidigal Filho, P .S., Sagrilo, E. & Mora, F. (2005). Adubações orgânica e mineral, produtividade do milho e características físicas e químicas de um Argissolo Vermelho-Amarelo. *Acta Scientiarum Agronomy*, *27,* 521-529.

Gong, W., Yan, X., Wang, J., Hu, T. & Gong, Y. (2009). Long-term manure and fertilizer effects on soil organic matter fractions and microbes under a wheat–maize cropping system in northern China. *Geoderma, 149*, 318-324.

Green, V. S., Stott, D. E., Cruz, D. E. & Curi, N. (2007). Tillage impacts on soil biological activity and aggregation in a Brazilian Cerrado Oxisol. *Soil & Tillage Research*, *92*, 114-121.

Hutchinson, J. J., Campbell, C. A. & Desjardins, R. L. (2007). Some perspectives on carbon sequestration in agriculture. *Agricultural and Forest Meteorology, 142*, 288-302.

IBGE (Instituto Brasileiro de Geografia e Estatística). Censo Agropecuário 2006: resultados preliminares. Rio de Janeiro: *Instituto Brasileiro de Geografia e Estatística*, 2006.

Intergovernmental Panel On Climate Change (Ipcc). (1996). Climate Change 1995. Working group 1. IPCC, *Cambridge University Press*, *Cambridge*, UK.

Iqbal, J., Hu, R., Lin, S., Hatano, R., Feng, M., Lu, L., Ahamadou, B. & Du, L. (2009). CO2 emission in a subtropical red paddy soil (Ultisol) as affected by straw and N-fertilizer applications: A case study in Southern China. *Agriculture, Ecosystems and Environment*, *131*, 292-302.

Jacobs, A., Rauber, R. & Ludwig, B. (2009). Impact of reduced tillage on carbon and nitrogen storage of two Haplic Luvisols after 40 years. *Soil & Tillage Research, 102*, 158-164.

Kukal, S. S., Rasool, R. & Benbi, D. K. (2009). Soil organic carbon sequestration in relation to organic and inorganic fertilization in rice–wheat and maize–wheat systems. *Soil & Tillage Research*, *102*, 87-92.

Lal, R. (1997). Residue management, conservation tillage and soil restoration for mitigating greenhouse effect by CO_2-enrichment. *Soil & Tillage Research*, *43*, 81-107.

Lal, R. (2003). Global potential of soil carbon sequestration to mitigate the greenhouse effect. *Critical Reviews in Plant Sciences*, *22*, 151-184.

Lal, R. (2004). Soil carbon sequestration impacts on global climate change and food security, *Science, 304*, 1623-1627.

Lal, R. (2008). Black and buried carbons' impacts on soil quality and ecosystem services. *Soil & Tillage research*, *99*, 1-3.

Lehmann, J., Gaunt, J. & Rondom, M. (2006). Bio-char sequestration in terrestrial ecosystems – a review. *Mitigation and Adaptation Strategies for Global Change, 11*, 403-427.

Leite, L. F. C., Doraiswamy, P. C., Causarano, H. J., Gollany, H. T., Milak, S. & Mendonca, E. S. (2009). Modeling organic carbon dynamics under no-tillage and plowed systems in tropical soils of Brazil using CQESTR. *Soil Tillage & Research*, *102*, 118-125.

Leite, L. F. C., Mendonça, E. S., Machado, P. L. O. A. & Matos, E. S. (2003). Total C and N storage and organic C pools of a Red-Yellow Podzolic under conventional and no tillage at the Atlantic Forest Zone, Southeastern Brazil. *Australian Journal of Soil Research, 41*, 717-730.

Leite, L. F. C., Mendonça, E. S., Machado, P. L. O. A., Fernandes-Filho, E. I. & Neves, J. C. L. (2004). Simulating trends in soil organic carbon of an Acrisol under no-till and disc-plow systems using the Century model. *Geoderma, 120*, 183-195.

Li, X. G., Li, Y. K., Li, F. M., Ma, Q., Zhang, P. L. & Yin, P. (2009). Changes in soil organic carbon, nutrients and aggregation after conversion of native desert soil into irrigated arable land. *Soil & Tillage Research, 104*, 263-269.

Maia, S. M. F., Ogle, S. M., Cerri, C. E. P. & Cerri, C. C. (2009). Effect of grassland management on soil carbon sequestration in Rondônia and Mato Grosso states, Brazil. *Georderma, 149*, 84-91.

Melero, S., López-Garrido, R., Murillo, J. M. & Moreno, F. (2009). Conservation tillage: Short- and long-term effects on soil carbon fractions and enzymatic activities under Mediterranean conditions. *Soil & Tillage research, 104*, 292-298.

Metay, A., Moreira, J. A. A., Bernoux, M., Boyer, T., Douzet, J. M., Feigl, B., Feller, C., Maraux, F., Oliver, R. & Scopel, E. (2006). Storage and forms of organic carbon in a no-tillage under cover crop system on clayey Oxisol in dryland rice production (Cerrados, Brazil). *Soil & Till. Research, 94, 122-132.*

Moraes, J. F. L., Neill, C., Volkoff, B., Cerri, C. C., Melillo, J., Lima, V. C. & Steudler, P. A. (2002). Soil carbon and nitrogen stocks following forest conversion to pasture in the Western Brazilian Amazon Basin. *Acta Scientiarum Agronomy, 24*, 1369-1376.

Nieder, R. & Benbi, D. K. (2008). *Carbon and Nitrogen in the Terrestrial Environment.* Springer Netherlands. 430p.

Oliveira, G. C., Dias-Junior, M. S., Resck, D. V. S. & Curi, N. (2004). Caracterização química e físico-hídrica de um latossolo vermelho após vinte anos de manejo e cultivo do solo. *Revista Brasileira de Ciência do Solo, 28*, 327-336.

Puget, P. & Lal, R. (2005). Soil organic carbon and nitrogen in a Mollisol in central Ohio as affected by tillage and land use. *Soil & Tillage Research, 80*, 201-213.

Purakayastha, T. J., Rudrappa, L., Singh, D., Swarup, A. & Bhadraray, S. (2008). Long-term impact of fertilizers on soil organic carbon pools and sequestration rates in maize–wheat–cowpea cropping system. *Geoderma, 144*, 370-378.

Roscoe, R. & Buurman, P. (2003). Tillage effects on soil organic matter in density fractions of a Cerrado Oxisol. *Soil & Tillage Research, 70*, 107-119.

Sa, J. C. M. & Lal, R. (2009). Stratification ratio of soil organic matter pools as an indicator of carbon sequestration in a tillage chronosequence on a Brazilian Oxisol. *Soil & Tillage Research, 103*, 46-56.

Sa, J. C. M., Cerri, C. C., Dick, W. A., Lal, R., Venske Filho, S. P., Piccolo, M. C. & Feigl, B. E. (2001). Organic matter dynamics and carbon sequestration rates for a tillage chronosequence in a Brazilian Oxisol. *Soil Science Society of America Journal, 65*, 1486-1499.

Sainju, U. M., Senwo, Z. N., Nyakatawa, E. Z., Tazisong, I. A. & Reddy, K. C. (2008). Soil carbon and nitrogen sequestration as affected by long-term tillage, cropping systems, and nitrogen fertilizer sources. *Agriculture, Ecosystems and Environment, 127*, 234-240.

Six, J., Elliott, E. T. & Paustian, K. (2000). Soil macroaggregate turnover and microaggregate formation: a mechanism for C sequestration under no tillage agriculture. *Soil Biology & Biochemistry*, *32*, 2099-2103.

Sodhi, G. P. S., Beri, V. & Benbi, D. K. (2009). Soil aggregation and distribution of carbon and nitrogen in different fractions under long-term application of compost in rice–wheat system. *Soil & Tillage Research, 103*, 412-418.

Spargo, J. T., Alley, M. M., Follet, R. F. & Wallace, J. V. (2008). Soil carbon sequestration with continuous no-till management of grain cropping systems in the Virginia coastal plain. *Soil & Tillage Research, 100*, 133-140.

Steiner, C., Teixeira, W. G. Lehmann, J., Nehls, T., Macêdo, J. L. V., Blum, W. E. H. & Zech, W. (2007). Long term effects of manure, charcoal and mineral fertilization on crop production and fertility on a highly weathered Central Amazonian upland soil. *Plant Soil, 291*, 275-290.

Ussiri, D. A. N. & Lal, R. (2009). Long-term tillage effects on soil carbon storage and carbon dioxide emissions in continuous corn cropping system from an alfisol in Ohio. *Soil & Tillage Research, 104*, 39-47.

West, T. O. & Post, W. M. (2002). Soil organic carbon sequestration rates by tillage and crop rotation. *Soil Science Society of America Journal, 66*, 1930-1946.

Yang, X. M., Drury, C. F., Wander, M. M. & Kay, B. D. (2008). Evaluating the effect of tillage on carbon sequestration using the minimum detectable difference concept. *Pedosphere, 18*, 421-430.

Zanatta, J. A., Bayer, C., Dieckow, J., Vieira, F. C. B. & Mielniczuk, J. (2007). Soil organic carbon accumulation and carbon costs related to tillage, cropping systems and nitrogen fertilization in a subtropical Acrisol. *Soil & Tillage Research, 94*, 510-519.

In: No-Till Farming
Editor: Earl T. Nardali
ISBN: 978-1-60741-402-5

Chapter 7

MAGNITUDE AND SEVERITY OF CHANNEL IRRIGATION-INDUCED LANDSLIDES ON DIFFERENT CULTIVATION SYSTEMS IN THE VEGETABLE GROWING AREA OF THE ULUGURU MOUNTAINS, TANZANIA

***D. N. Kimaro*[a], *J. Poesen*[b], *J. A. Deckers*[c] *and H. B. Msita*[a]**

[a]Department of Agricultural Engineering and Land Planning, Sokoine University of Agriculture, P.O. Box 3003, Morogoro, Tanzania

[b]Physical and Regional Geography Research Group, Katholieke Universiteit Leuven, Geo-Institute, Celestijnenlaan 200 E, B-3001 Heverlee, Belgium

[c]Division of Soil and Water Management, Geo-Institute, Katholieke Universiteit Leuven, Celestijnenlaan 200 E, B-3001 Heverlee, Belgium

ABSTRACT

A survey was carried out to determine the effects of channel irrigation on land degradation in the vegetable growing area on the northern slopes of the Uluguru Mountains, Tanzania. Sixteen irrigation-induced landslides were characterised in terms of their dimensions, associated geomorphic characteristics, landslide typology and morphological and physical characteristics of the soils. Qualitative assessment and descriptive statistics were widely employed in the exploratory analysis of the collected data and related factors. This included the estimation of means and standard deviations of some critical variables of the study. Where applicable, Least Squares Means (LSMean) was used to compare the treatment means at 5 percent level of probability. From the results, it was observed that most of the landslides occurred on debris slopes on the linear/concave slope complex as earth flows and or earth slides with few occurring on the amphitheatres. The dominant slide scar angles ranged from 22-35°. Channel irrigation in this area provoked relatively small sized landslides with maximum width and depth ranging from 7 to 19 m and 0.3 to 0.9 m respectively while their length ranged from 6 to 70 m. The mean damaged cropland area ranged from 211 to 1327m^2 while volume of soil loss ranged from 36 to 72m^3. Both the damaged cropland area by landslides and volume of soil loss due to the latter are the major indicators of land degradation due to channel irrigation and these parameters are proposed to be one of the yardsticks in determining the sustainability of production of vegetables in the study area. Damaged cropland area was more severe on flat cultivation system compared to ridge and contour cultivation systems ($P \leq 0.05$). This phenomenon is explained in terms of loss of crop yield and reduced farmers income. The volume of soil loss followed the trend flat cultivation > contour cultivation > ridge cultivation though the values were not statistically significantly different.

Keywords: Irrigation-induced landslides, cultivation system, Uluguru Mountains, Tanzania

1.0. Introduction

1.1. Landslide Characteristics

Landslide erosion occurs on steep slopes and involves transfer of soil materials from higher slopes to lower grounds under the influence of gravity (Lal, 1990). Varnes (1958) used the term landslide to describe downward and outward movement of slope forming materials composed of rock, soils, artificial fills or a combination of these materials. In a later publication Varnes (1978) changed the term landslide to "slope movement" which includes also slower movements such as creep and solifluction. Landslides can be categorised into 3 major types namely falls, slides and flows (Varnes, 1978). These categories are based on the physical state of materials removed. Falls are defined as the falling through the air down a cliff or the bouncing of individual blocks of rock or soil down a steep slope (Varnes, 1978). Slides involve a rapid movement of mass of rock or debris gliding on one or several slide planes on a substratum and creating a considerable friction along the slide scar (Temple and Rapp, 1972; Varnes, 1978). Flows are movement by plastic flow or by shear along many planes in masses of soil or debris (Lundgren, 1978; Varnes 1978; Zinck, 1990). Furthermore, slides have been categorised based on the type of movement involved, as translational and rotational landslides (Larsson, 1989; Lopez and Zinck, 1991; Reid and Dunne, 1996). The translational landslides are those involving shear failure on a fairly plane surface running roughly parallel to the general slope of the ground whereas rotational landslides takes place when a rapid shear failure occurs on a well-defined somewhat curved slip surface (Larsson, 1989).

Ahmad and McCalpin (1999) identified two forms of landslides in Kingston Metropolitan area, Jamaica: deep-seated which is associated with bedrock and shallow landslides which occur in colluvium. According to the authors deep-seated landslides represent failure of deeply weathered bedrock material whereas the shallow landslides typically are associated with a failure of residual or transported mantle also known as residual colluvium. According to Varnes (1978) classification the famous landslides which occurred in the Uluguru Mountains in Tanzania in 1970 (Temple and Rapp, 1972) fall under the category of debris slides. The typical slide scars were 5-10 m wide, 10-50 m long and up to 2 m deep and wide near the upper end, narrowing in a wedge-like fashion downslope (Temple and Rapp, 1972; Lundgren, 1978).

1.2. Factors Influencing the Occurrence of Landslides

The triggering of landslides depends on several complex but interrelated variables such as prolonged heavy rainfall (Temple and Rapp, 1972; Larsson, 1989), soil characteristics (Westerberg, 1999; Mburu, 2001) slope characteristics (Lopez and Zinck, 1991), lithology and geomorphology (Ahmad and McCalpin, 1999, Westerberg and Christiansson, 1999), and vegetation, land use and infrastructure (Larsen and Torres-Sanchez, 1998).

Rainfall is one of the important factors considered to have an influence on the occurrence of landslides. Larsen and Simon (1993) reported that 40 out of 42 storms recorded in Puerto Rico between the year 1959 and 1990 had triggered tens to hundreds landslides each. Larsson (1989) observed high concentration of landslides activity in the Nyandarua range in Kenya during the long rains of 1978, 1981 and 1986. Prolonged heavy rainfall which occurred at Mgeta in the Uluguru Mountains, Tanzania in 1970 resulted in over 1,000 landslides (Temple and Rapp, 1972). These catastrophic events were triggered by high rainfall intensity amounting to 100.7 and 185 mm in three and 72-hours respectively. Similar observations on the contribution of high rainfall intensity areas were made in the Serra das Araras in Brazil (Rapp and Nyberg, 1991) and in south eastern France (Julian and Anthony, 1996). The high pore water pressure due to heavy rainfall (Johnson and Sitar, 1990) in combination with other factors such as steep slopes and favourable soil characteristics are reported to be instrumental in triggering landslides (Brookes and Anderson, 1995).

Soil physical characteristics largely determine the occurrence of landslides. A study conducted in Columbia on the factors conditioning the occurrence of landslides by Lopez and Zinck (1990) show that soil physical properties such as soil texture, bulk density, water retention and soil consistence limits were key factors influencing the occurrence of landslides. Larsen and Torres-Sanchez (1996) while mapping landslides distribution hazards in Puerto Rico observed high landslide frequencies on soils mainly Ultisols and Inceptisols developed on steep slopes and ridge tops. A study conducted in a tropical mountainous area in Ecuador by Vanacker (2002) reported that rotational and translational slides were more related to an increased soil moisture and soil pore water pressure. The shallow debris slides occurred in Nyandarua range in Kenya was mainly caused by the failure surfaces that have developed in the transition zone between soil solum and the bedrock (Larsson, 1989; Mburu, 2001). The authors observed that the larger and deeper landslides were situated in clay rich Andisols and Nitosols, often with very deep soil profiles and increasing clay content with depth.

The degree, length and shape of slope are the main slope characteristics reported to have potential influence in the development and occurrence of landslides (Larsson, 1989). Steep slopes are more vulnerable to landslides (Lopez and Zinck, 1991). In Kingston Metropolitan area in Jamaica, Ahmad and McCalpin (1999) observed that shallow landslides were more common on slopes between 20° and 35°. According to the authors, only few landslides were observed on slopes steeper than 35°. The low landslide densities observed on these steep slopes were attributed to absence of colluvium or shallow soils. Larsen and Torres-Sanchez (1998) working in Puerto Rico found that slope angles exceeding 12° had the greatest frequency of landslides. In northern Thailand, the occurrence of translational slides (shallow landslides) was more common on slopes varying from 70% to 120% (Turkelboom, 1999). The reported landslides in Mgeta area in the Uluguru Mountains, Tanzania, (Temple and Rapp, 1972) were restricted to slopes steeper than 26° and were most common on slopes steeper than 32°.

Lithology, faults, and downslope curvature are also reported to be among the important factors controlling the occurrence of landslides (Ahmad and McCalpin 1999). According to Westerberg (1999), the highlands of East Africa characterised by complex of concave slopes are prone to slope instability and that mass movements are the prime contemporary denudation process. The microtopography of slopes is one of the factors responsible for triggering landslides and mudflows (Fernandes *et al.,* 1994; Westerberg and Christiansson, 1999). According to the authors troughs which exist between rock outcrops and soil-filled concavities and hillslope hollows act as water reservoirs that influence the occurrence of mass movement. Westerberg and Christiansson (1999) reported that linear down cutting of pre-existing drainage lines by extreme fluvial incision and fluvial removal of colluvium result in steep morphology which is commonly associated with slope instability. The authors attribute landslides to the presence of geological structures that control the development of weak layers serving as potential sliding planes. Also, rapid weathering produces regolith prone to high pore water pressure (Westerberg, 1999) which according to Temple and Rapp (1972) cause water blowouts and hence the occurrence of landslides.

Several studies have shown the importance of vegetation cover and land use in influencing the occurrence of landslides (Dhakal *et al.*, 1999; Luckman *et al.*, 1999; Baeza and Corominas, 2001). In tropical mountainous area in Ecuador, Vancker (2002) noted that with increasing deforestation, seepage forces became an important factor in inducing the development of shallow landslides. The study conducted by Larsen and Torres-Sanchez (1998) in Puerto Rico showed that the average frequency of landslides increased markedly as land use changed from forest to agriculture to construction of roads and structures such as terraces and irrigation channels. In this study, the authors noted that the number of landslides occurring in connection with roads and other structures on average exceeded the number of landslides in forest by 2.3 to 8.4 times. In southern France, Julian and Anthony (1996) reported that deforestation, quarrying, urbanization and road network developments has lead to the destabilization of the coastal hillslopes hence favouring the development of numerous shallow landslides. Of the more than 100 landslides that occurred in Mgeta area Tanzania in 1970, 47% originated in cultivated fields, 46% in grassland and less than 1% in woodland covered areas (Temple and Rapp, 1972).

1.3. Assessing the Severity of Landslides Erosion

The severity of landslides is commonly estimated in terms of: (1) volume of soil loss (Temple and Rapp, 1972), (2) frequency of occurrence, (3) total affected area and the susceptibility of the area to landslide erosion (Lopez and Zinck, 1991). Ahmad and McCalpin (1999) proposed a rating system for mapping landslides severity on the basis of landslide density as percentage of total slide area/total land unit area. Temple and Rapp (1972), working in the Uluguru Mountains, Tanzania developed a model to estimate the volume of soil loss from landslide scars which is expressed as shown in equation (1) below:

$$V = \pi ab \sin\theta \, (0.21a + 0.146L) \qquad (1)$$

Where: V = Volume of soil loss (m^3)
a = Maximum width of the eroded area (m)
b = Maximum depth of cutting measured vertically (m)
θ = The angle between the slope and the vertical depth
L = The total measured length of the eroded slope (m).

1.4. Irrigation Induced Landslides in Tanzania

Smallholder irrigation system is typical of many mountainous areas in Tanzania. Using a wide range of local technology, farmers in these areas have managed to increase the production on their land through irrigation. Off-season agriculture in these areas involves the production of high value crops notably vegetables.

In the Uluguru Mountains, vegetable production is mainly done under channel irrigation system but on small farm plots. It is an important and dependable source of food and cash to meet both household requirements and the market demand of a number of towns in Tanzania (Kilasara and Rutatora, 1993; Lyamuya *et al.*, 1994; Verbesselt and Mertens, 2000). However vegetable production in this area is faced with multiple problems including land shortage to which landslide is a contributing factor. Landslides research in the Uluguru Mountains is comprehensive and it has been mainly concerned with naturally occurring landslides which coincidentally is of big magnitude (Temple and Rapp, 1972; Lundgren, 1978; Westerberg, 1999; Westerberg and Christiansson, 1999). Channel irrigation induced landslides are of primary importance when land scarcity is a constraint to development as it is in the northern slopes of the Uluguru Mountains. Studies on man-made landslides are rare not only in the Uluguru Mountains but in Tanzania as whole. Information on when and where such landslides occur and their effect on land management are still unknown in Tanzania. Equally the direct and indirect economic consequences of the landslides caused by irrigation are not fully known (Kongola, 2000). Therefore, both onsite and offsite effects due to irrigation provoked landslides should form part and parcel of an integrated soil and water conservation interventions in mountainous zones. The urgent need to assess irrigation induced soil erosion in the Uluguru Mountains is inevitable taking into account the rapid expansion of the vegetable industry whose major input is irrigation. The objective of this study was to assess the magnitude and severity of channel irrigation induced landslides on different cultivation systems in the vegetable growing area of the Uluguru Mountains, Tanzania.

2.0. Description of the Study Area

The study area is located in the northern slopes of the Uluguru Mountains, Tanzania. It is concentrated to an area outlined by 350295E and 354368E and 9237500N and 9243697N UTM coordinates (Figure 1). The size of the study area was roughly 100 ha. This is the area mainly used for vegetable production under channel irrigation system with many farmers reporting incidences of land sliding during off-season cultivation. The cropping system is closely related to the geomorphic characteristics of the study area (Figure 2). About 10% of the mountain debris slopes are used for vegetable production like carrots, spinach, salads, onions, leaks, cabbage, celery and sprouting broccoli. Vegetables are grown in this zone during dry season under channel and drag horse sprinkler irrigation system. The higher mountain ridge slopes are not cultivated during high rainfall intensity especially in the month of April attributed to the fact that farmers are attempting to avoid rain-induced landslides during wetter periods. Maize is also grown in this unit along with vegetables in the months of October to March (short rains maize). The remaining area of the mountain ridges is mainly used for cultivation of pigeon peas, beans and few scattered banana crops and seasonal fallow plots. The talus slopes are mainly used for banana cultivation intercropped with maize while more than 80% of the V-shaped valleys and landslide scars are cultivated with banana as the main crop. The ridge and hill summits are mainly under settlements.

In this area agricultural land use has changed dramatically over the last 20 years. Before 1980's only few crops such as maize, beans, sweet potato and cassava were grown (Verbesselt and Mertens, 2000).

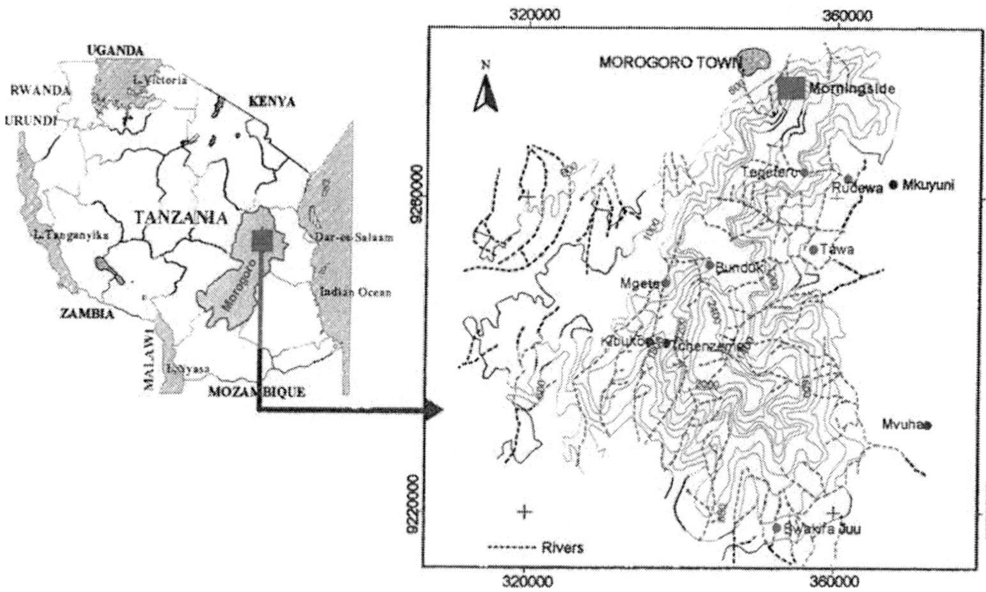

Figure 1. Location of the study area

A common rotation was two to three years followed by a fallow period of three to four years. Temperate vegetables were cultivated on a small scale (Kimaro *et al.*, 1999). After 1980's the intensification of cropping systems increased due to population pressure, land scarcity and expansion of the nearby Morogoro municipality and Dar es Salaam city (Mwango, 2000; Kimaro *et al.*, 2003). Cultivation of vegetables increased to meet the demand of the growing population in the nearby towns. Fallow cultivation and crop rotation was discouraged and continuous cultivation intensified. At present vegetables can be grown anywhere and almost independent of the soils. Mixtures of subsistence crops are grown in most of the units. Maize, beans, pigeon peas and cassava are exclusively grown for home consumption. Vegetables and bananas are mainly grown for sale. Small scale intercropping occurs in almost every field. The present land use shows clearly that every piece of the land is cultivated and cropping systems are highly variable and complex (Kimaro *et al.*, 1999). Crops are mainly grown as flat cultivation system apart from vegetables which are grown in narrow and broad band ridges in almost all geopedologic units. Most farmers remove all the residues and/or set their fields on fire before cultivation.

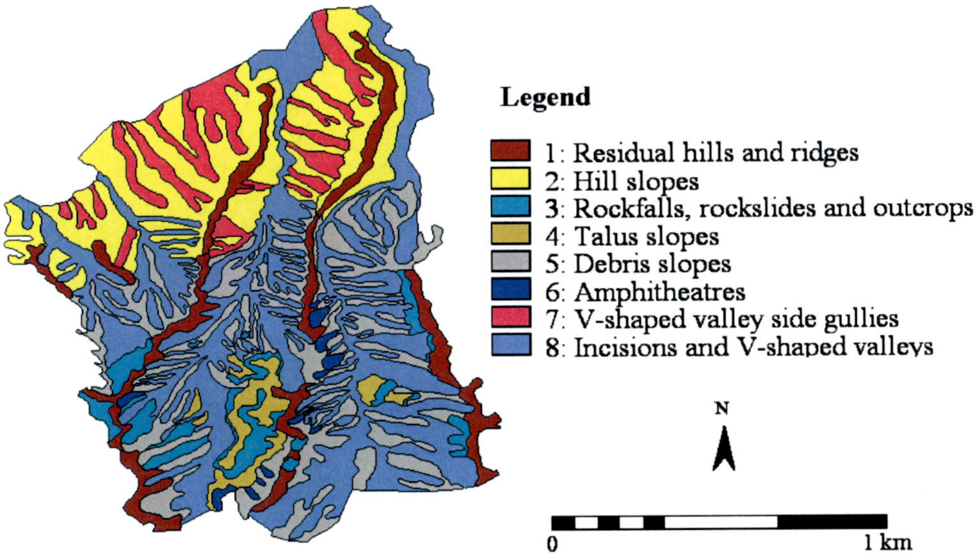

Figure 2. Geomorphic characteristics of the study area in the northern slopes of the Uluguru Mountains, Tanzania

3.0. Survey and Data Collection

The survey and data collection was carried out during May to November 1999 and 2000 respectively. Inventory of landslides scars was made by rapid survey through visual inspection in the farmers' fields and interviewing. For each landslide scar, notes were made on scar dimensions and landslide typology described according to Varnes, (1978) classification system: period of movement, geomorphic characteristics including type of geoform, slope form and slope angle. Other information collected included morphological and physical characteristics of the soils and land management systems. Informal interviews were held with farmers to establish when land sliding occurred, impact on farming practices, land use management and cultivation systems, yields, prices and the damage caused in the immediate surroundings. In addition soil profile descriptions were made at each landslide scar.

Dimensions of landslide scars (width, depth and length) were measured using measuring tape and clinometer (Temple and Rapp, 1972; Lundgren, 1978; Westerberg, 1999). These dimensions were used to calculate volume of soil loss from the landslide scars using the model suggested by Temple and Rapp (1972) as follows:

$$V = \pi ab \sin\theta (0.021a + 0.146L)$$

Where: V = volume (m3)
a = maximum width of the eroded area (m)
b = maximum depth of cutting measured vertically (m)
θ = maximum angle between the slope and the vertical depth
L = the total measured length of the eroded slope (m).

Soil profile morphological characteristics were described according to FAO (1990). Medium and coarse pores were counted visually on the landslide scars and recorded as number per unit area of a square decimetre (FAO, 1990). Bulk density was determined using undisturbed samples according to core sample method (Blake and Hartage, 1986). Soil moisture retention characteristics were studied using pressure plate extractor method (Klute, 1986). Particle size distribution was done using standard hydrometer method (Gee and Bauder, 1986) after destroying soil organic matter to obtain sand, silt and clay fractions.

Qualitative assessment and descriptive statistics were widely employed in the exploratory analysis of the collected data and related factors. This included the estimation of means and standard deviations of some critical variables of the study. The effect of multiple factors were statistically analysed according to the General Linear Model (GLM) of the Statistical Analysis System (SAS, 1999). Where appropriate, Least Squares Means (LSMean) was used to compare the treatment means at 5 percent level of probability.

4.0. Characteristics of Channel Irrigation-Induced Landslides

4.1. Soils, Landform and Associated Landslide Morphological Characteristics

Characteristics of the studied channel irrigation-induced landslides scars are presented in Table 1. The spatial distribution of the studied irrigation-induced landslides is presented in Figure 3. In total, 16 landslide scars were identified over a seven km of irrigation channel. All the studied slide scars occurred on debris slopes and amphitheatres below an irrigation channel and on incisions and V-shaped valleys. These geomorphic units are characterised by linear-concave slope complexes and are mostly used for vegetable production during off-season crop normally between the months of May/June through October/November. About 44% of them occurred on debris slopes, 25% on the amphitheatres and 31% on incisions and V-shaped valleys. Channel irrigation in this area provoked relatively small sized landslides. The dominant slide scar angles ranged from 22° to 35°. As shown in Table 1 the length of most slides scars ranged from 5 m to 70 m while width and depth ranged from 7 m to 19 m and 0.3 m to 0.9 m respectively. The observed slide scars were wide near the upper end, narrowing in a wedge-like fashion down slope. According to Varnes (1978) classification system these types of landslides involve earth slides and earth flows.

The dominant soil types in the study area have been classified accoding to FAO *et al.* (1998) as *Lithic Leptosols, Epileptic Cambisols, Endoleptic Cambisols and Endoskeletic Cambisols* (Table 2). The *Leptic Cambisols* and *Epileptic Cambisols* are associated with debris slopes and overly a hard rock at shallow depth.

Table 1. Characteristics of irrigation-induced landslide scars on different geomorphic units and cultivation systems

Cultivation system	Slide no.	Geoform	Slope form	Slope angle	Type of movement	Scar dimensions			Month-Year of movement
						Maximum width (m)	Maximum depth (m)	Length(m)	
Flat cultivation	LIR4	Debris slope	Straight/concave	31-35°	earth slide	12.6	0.65	5.9	July 1999
	LIR5	Debris slope	Straight/concave	31-35°	earth flow	19.2	0.25	70.0	Sept. 1999
	LIR13	Debris slope	Straight/concave	22-31°	earth slide	11.2	0.30	14.1	Aug. 1999
Ridge cultivation	LIR8	Debris slope	Straight	22-31°	earth flow	11.5	0.50	9.9	July 1999
	LIR11	Debris slope	Straight/concave	22-31°	earth slide	11.2	0.24	16.3	Aug. 1999
Contour cultivation	LIR2	Amphitheatre	concave	22-31°	earth flow	7.9	0.63	12.4	July 1999
	LIR14	Amphitheatre	concave	22-31°	earth flow	8.9	0.70	9.4	Aug. 1999
Flat cultivation	LIR6	Debris slope	Straight/concave	31-35°	earth flow	11.9	0.28	50.0	Nov. 2000
	LIR7	Debris slope	Straight/concave	22-31°	earth flow	8.5	0.30	22.0	July 2000
	LIR16	Debris slope	Straight/concave	22-31°	earth flow	8.4	0.35	20.0	Oct. 2000
Ridge cultivation	LIR3	Debris slope	Straight/convex	22-31°	earth flow	8.7	0.77	8.7	June 2000
	LIR10	Debris slope	Straight/concave	22-31°	earth flow	8.7	0.9	11.8	May 2000
Contour cultivation	LIR1	Amphitheatre	concave	22-31°	earth flow	8.5	0.70	8.6	Aug. 200
	LIR9	Amphitheatre	concave	22-31°	earth flow	6.3	0.9	7.9	July 2000
	LIR12	Amphitheatre	concave	22-31°	earth flow	8.2	0.6	6.6	May 2000
	LIR15	Amphitheatre	concave	22-31°	earth flow	7.5	0.65	12.9	Sept 2000

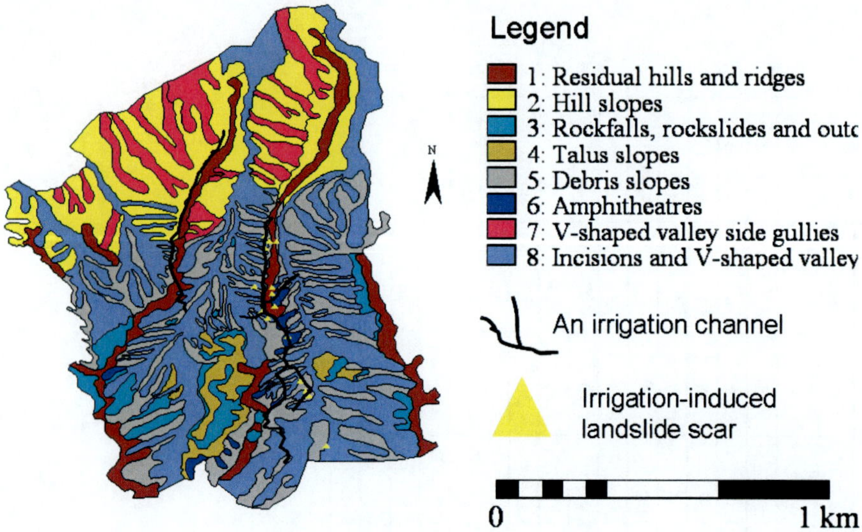

Figure 3. Spatial distribution of irrigation-induced landslides on different geomorphic units along the main irrigation channel in the study area

Table 2. Soil morphological characteristics of some selected irrigation-induced landslide scars in different geomorphic units

Cultivation system	Slide no.	Geoform	Soil type	Horizon	Depth (cm)	Nature of the horizon
Flat cultivation system	LIR5	Debris slope	Lithic Leptosols	Ap	0-25	grSL
				R	25+	hard rock
	LIR16	Debris slope	Epileptic Cambisols	Ap	0-35	grSL
				R	35+	hard rock
Ridge cultivation system	LIR8	Debris slope	Epileptic Cambisols	Ap	0-50	grSCL
				R	50+	hard rock
	LIR3	Debris slope	Epileptic Cambisols	Ap	0-77	vgrSCL
				R	77+	hard rock
Contour cultivation system	LIR14	Amphitheatre	Endoleptic Cambisols	Ap	0-70	grSL
				C	70+	SL saprolite
	LIR9	Amphitheatre	Endoleptic Cambisols	Ap	0-90	grSL
				C	90+	SCL saprolite

grSL = gravely sandy loam, grSCL = gravely sandy clay loam, vgrSCL = very gravely sandy clay loam, SCL = sandy clay loam, SL = sandy loam

The *Endoskeletic Cambisols* occur mainly on the amphitheatres overlying a saprolite below a depth of 70 cm from the soil surface. On the amphitheatres contour cultivation system is practiced while flat and ridge cultivation is mainly done on debris slopes. The top of the hard rock and saprolite constitutes the bottom of the landslides along which slopes have been undermined and hence these surfaces formed the sliding plane of the studied landslide scars. In Table 3, some physical characteristics of the soils above the stable surface of some selected landslide scars on different cultivation systems are presented. From Table 3, it appears that the bulk density of the soils above the stable surface is very low ranging from 1.1 g/cm^3 to 1.3 g/cm^3 with very high total porosity ranging from 39% to 52%. Medium and coarse pores per dm^2 are high ranging from 6 to 13 (FAO, 1990) while surface water content at field capacity ranges from 10.2 to 16.2 percent volume. These characteristics indicate favourable conditions for subsurface seepage from irrigation channels which could eventually results in sudden rises in pore water pressure that cause a reduction in shear strength below the force of the applied shear stress and thus causing a shear failure (Temple and Rapp, 1972; Westerberg, 1999). Therefore, the occurrence of channel irrigation-induced landslides observed in the study area could be attributed to high pore water pressure caused by water seeping from irrigation channel through subsurface seepage and piping through medium and course pores (Table 3). The phenomenon is further explained by low bulk density and high porosity both fine, medium and course pores of the soil material above the stable surface which are critical conditions favouring high percolation rate of excess water (overland flow) from irrigation and vegetable fields coupled with slope concavities with converging water-flow (Turkelboom, 1999).

Table 3. Physical characteristics of the soils above the stable surface of some selected irrigation-induced landslide scars

Cultivation system	Slide no.	Total Porosity (%)	Medium/coarse pores (No/dm^2)	Bulk density (gcm^{-3})	Surface water content at field capacity (%volume)
Flat cultivation system	LIR5	51.8	12	1.1	11.2
	LIR16	50.8	13	1.1	10.2
Ridge cultivation system	LIR8	38.6	9	1.3	13.8
	LIR3	43.2	6	1.2	13.2
Contour cultivation system	LIR14	46.0	13	1.2	14.3
	LIR9	48.4	13	1.2	16.2

4.2. Effects of Channel-Induced Irrigation Landslides

Damaged cropland area and volume of soil loss are the major onsite effects of channel irrigation-induced landslides in the study area (Table 4). Irrigation-induced landslides in the Uluguru Mountains therefore pose a serious threat to the economy of the household. The mean damaged cropland area ranged from 211 m^2 to 1327 m^2 for the year 1999 and 2000 off-season crop while volume of soil loss ranged from 36 m^3 to 72 m^3 (Table 4). Both damaged cropland area and volume of soil loss was more severe on flat cultivation compared to ridge and contour cultivation systems ($P \leq 0.05$). This could be attributed to poor soil conservation measures on flat cultivation system and shallow soils overlying a hard rock (Table 2). The volume of soil loss followed the trend flat cultivation > contour cultivation > ridge cultivation though the values were not significantly different. These results point out the importance of channel irrigation-induced landslides particularly on reduction of cropland area and land degradation as major constraint to development in the Uluguru Mountains.

The effects of channel irrigation-induced landslides are further explained by loss of crop yield and reduction of farmers' income (Table 5). The landslides caused a loss of vegetables cultivated on the collapsed land and the surrounding fields. The mean yield loss of vegetables ranged from 70 kg/ha on contour cultivation system to 445 kg/ha on flat cultivation system which correspond to an income loss of 10,500/= Tsh/ha to 68,250/= Tsh/ha respectively (Table 5). The income loss followed a similar trend to that of the damaged cropland area where as income loss on flat cultivation was highest at 46% followed by ridge cultivation 11% and least on contour cultivation with only 7%. It should be realised that channel irrigation-induced landslides have also resulted in the long-term setting-aside of the land as a potential risk area for the occurrence of future landslides. The uphill area of the fields is at great risk of retrogressive sliding owing to the very steep slide scarp while the irrigation channel is exposed to the risk of collapsing.

Table 4. Influence of type of cultivation on cropland damage by irrigation-induced landslides

System of cultivation	Damaged cropland area (m^2)	Volume of soil loss (m^3)
Flat cultivation system	1327[a]	72[a]
Ridge cultivation system	238[b]	36[a]
Contour cultivation system	211[b]	45[a]
Standard error	±221	±18

Means with the same superscripts within columns are not significantly (P≤0.05) different

Table 5. Effect of channel irrigation-induced landslides on vegetable yield and farmers' income on different cultivation systems

Land management system	Mean farm size	Damaged cropland area	Yield	Yield loss	Net income	Income loss	Income loss
	(ha)		kg/ha		Tsh/ha		%
Flat cultivation	0.28	0.13	980	455	147,000/=	68,250/=	46
Ridge cultivation	0.27	0.03	945	105	141,750/=	15,750/=	11
Contour cultivation	0.30	0.02	1050	70	157,500/=	10,500/=	7

1 US$ = TAS 1,025/=, 1 EURO = TAS 1,102/= (February 2003)

5.0. CONCLUSION

Irrigation induced landslides is one of the dominant form of land degradation on the ridges of the northern slopes of the Uluguru Mountains with slope gradient range of 47% to 120% where they have affected the major part of the land used for vegetable production.

Their distribution follows the trend: on the debris slopes (44%), on the incisions and V-shaped valleys (31%) and on the amphitheatres (25%). The adverse consequences of irrigation-induced landslides in the study area include significant reduction of crop yield and farmers' income attaining up to 50%.

The damaged cropland area and volume of soil loss due to channel irrigation-induced landslides were more severe on flat cultivation compared to ridge and contour cultivation systems. Both the damaged cropland area by channel irrigation-induced landslides and volume of soil loss due to the latter are the major indicators of land degradation due to channel irrigation and these parameters are proposed to be one of the yardsticks in determining the sustainability of production of vegetables in the study area.

Channel irrigation-induced landslides are of primary importance when land scarcity and degradation is a constraint to development as it is in the northern slopes of the Uluguru Mountains, Tanzania. Hence, it is an important form of land degradation and should be considered as a major constraint to agricultural development in the study area

ACKNOWLEDGMENTS

The authors wish to express their sincere appreciation to Soil and water Conservation on the Uluguru Mountains Research Project under SUA-FLEMISH Universities (SUA-VLIR) Collaboration Programme for the financial assistance towards this work. The faremers in the study area are highly acknowledged.

REFERENCES

Baeza, C. & Corominas, J. (2001). Assessment of shallow landslide susceptibility by means of multivariate statistical techniques. *Earth Surface Processes and landforms*, *26*, 1251-1263.

Blake, G. R. & Hartage, K. H. (1986). Bulk density. In: *Methods of Soil Analysis, Part 1 Physical and Mineralogical Methods. (edited by Klute, A.)* Number 9 in the series Agronomy, Madison, Wisconsin, pp 364-374.

Brooks, S. M. & Anderson, M. G. (1995). The determination of suction-controlled slope stability in humid temperate environments. *Geografiska Annaler*, *77A*, 11-22.

Dhakal, A. S., Amda, T. & Aniya, M. (1999). Landslide hazard mapping and the application of GIS in the Kulekhani watershed, Nepal. *Mountain Research and Development*, *19*, 3-16.

Fernandes, N. F., CoelhoNetto, A. L. & Lacerda, A. W. (1994). Sub-surface hydrology of layered colluvium mantels in unchannelled valleys-South-eastern Brazil. *Earth Surface Processes and Landforms*, *19*, 609-626.

Food and Agriculture Organisation (FAO). (1990). Guidelines for Soil Profile Description. 3rd edition. FAO, Rome, Italy. pp 66.

Food and Agriculture Organisation (FAO), International Soil Reference and Information Centre (ISRIC) and International Society of Soil Science (ISSS). (1998). *World*

Reference Base for Soil Resources, World Soil Resources Report 84, FAO, Rome, Italy. pp 88.

Gee, G. W. & Bauder, J. W. (1986). Particle size analysis: In: *Methods of Soil Analysis, Part1; Physical and Mineralogical Methods, 2 edition. (Edited by Klute, A.)* Number 9 in the series Agronomy, Madison, Wisconsin, pp 383-411.

Johnson, P. A. & Sitar, N. (1990). Hydrologic conditions leading to debris flow initiation. *Canadian Geotechnical Journal*, *27*, 789-801.

Julian, M. & Anthony, E. (1996). Aspects of landslide activity in the Mercantour Massif and the French Riviera, Southeastern France. *Geomorphology*, *15*, 275-289.

Kilasara, M. & Rutatora, D. F (1993). The socio-economic and land use factors affecting the land degradation of the Uluguru Catchment in Morogoro, Tanzania. In: *Agriculture and the Environment (edited by Rutachokozibwa, V., Rutatora, D.F., Lugeye, S.C. and Mollel, N.M.)* Sokoine University of Agriculture, Morogoro, Tanzania. pp 27-31.

Kimaro, D. N., Kilasara, M., Noah, S. G., Donald, G., Kajiru, K. & Deckers, J. A. (1999). Characteristics and management of soils located on specific landform units in the northern slopes of Uluguru Mountains, Tanzania. In, *Agricultural Research Challenges for the 21st Century. Proceedings of the Fourth Annual Research Conference of the Faculty of Agriculture*. (Edited by Faculty of Agriculture (FoA)) Sokoine University of Agriculture, Morogoro, Tanzania, November 17th-19th 1999, pp. 234-242.

Kimaro, D. N., Msanya, B. M., Kimbi, G. G., Kilasara, M., Deckers, J. A., Kileo, E. P. & Mwango, S. B. (2003). Computer captured expert knowledge for land evaluation of mountainous areas: A case study of Uluguru Mountains, Morogoro, Tanzania. Revised manuscript submitted for publication to Uniswa Journal of Agriculture, *Science and Technology*, pp. 20.

Klute, A. (1986). Water retention. In: *Methods of Soil Analysis, Part1; Physical and Mineralogical Methods, 2 edition. (Edited by Klute, A.)* Number 9 in the series Agronomy, Madison, Wisconsin, pp 635-660.

Kongola, M. J. M. (2000). *Performance evaluation of an indigenous irrigation system at Towero Village, Western Uluguru Mountains, Tanzania*. MSc thesis, Sokoine University of Agriculture, Morogoro, Tanzania. pp 154.

Lal, R. (1990). *Soil erosion in the tropics. Principles and management.* McCraw Hill. U. S. A. pp. 580.

Larsen, M. C. & Simon, A. (1993). A rainfall intensity-duration threshold for landslides in humid-tropical environment, Puerto Rico. *Geografiska Annalar*, *75A*, 13-23.

Larsen, M. C. & Torres-Sanches, A. J. (1998). The frequency and distribution of recent landslides in three montane tropical regions in of Puerto Rico. *Geomorphology*, *24*, 309-331.

Larsen, M. C. & Torres-Sanchez, A. J. (1996). Geographic relations of landslide distribution and assessment of landslide hazards in the Blanco, Cibuko and Coamo River Basins, Puerto Rico. U.S Geological Survey Water Resources Investigation Report 95-4029. pp. 56.

Larsson, M. (1989). Landslides in the mountain areas of Kenya: Comparative studies of different slopes within the Nyandarua Range. In: *Proceedings of the Third National Workshop* (Edited by Thomas, D. B., Biamah, B. K., Kilewe, A. M., Lundgren, L. & Muchoge, B.O.). 16-19 September 1986, Department of Agricultural Engineering,

University of Nairobi and Swedish International Development Authority (SIDA) Nairobi, Kenya. pp. 123-136.

Lopez, J. H. & Zinck, J. A. (1991). GIS-assisted modelling of soil-induced mass movement hazards: A case study of the upper Coello river basin, Tolima, Colombia. *ITC Journal*, 10991-4, 202-220.

Luckman, P. G., Gibson, R. D. & Derose, R. C. (1999). Landslide erosion risk to New Zealand pastoral steeplands productivity. *Land Degradation and Development*, *10*, 49-65.

Lundgren, L. (1978). Studies of soil and vegetation development on fresh landslide scars in the Mgeta Valley, Western Uluguru Mountains, Tanzania. *Geografiska annaler*, *60A* (3-4), 91-126.

Lyamuya, V. E., Noah, E. E., Kilasara, M., Kirenga, E. J. & Burgess, N. D. (1994). *Socio-economic and Land Use Factors Affecting the Degradation of the Uluguru Mountains Cathment in Morogoro, Region, Tanzania.* Regional Natural Resources Office: Morogoro, Tanzania. pp 38. 7 maps.

Mburu, D. M. (2001). Mass movement in agricultural land in the eastern slopes of Nyandarua Mountain ranges (Kenya) and the effect on agricultural production. Paper presented at the 19th Annual Conference of the Soil Science Society of East Africa, 2nd to 7th December 2001, Moshi, Tanzania. pp. 15.

Rapp, A., Li, J. & Nyberg, R. (1991). Mudflow disasters in mountainous areas. *Ambio*, *20*, 210-218.

Rapp, A., Axelsson, V., Berry, L. & Murray-Rust, D. H. (1972). Soil erosion and sediment transport in the Morogoro river catchment, Tanzania. *Geografiska Annaler, 54A*, (3-4), 125-155.

Reid, L. M. & Dunne, T. (1996). *Rapid evaluation of sediment budgets.* Catena Verlag, Reiskirchen, Germany. pp. 164.

SAS, (1999). Statistical Analysis System, SAS/STAT User's guide release. SAS Institute Inc, Carry NC, USA.

Temple, P. H. & Rapp, A. (1972). Landslides in the Mgeta area, Western Uluguru Mountains, Tanzania. Geomorphological effects of sudden heavy rainfall.

Turkelboom, F. (1999) *On-farm diagnosis of steepland erosion in northern Thailand: Integrating spatial scales with household strategies*. PhD thesis, Catholic University of Leuven, Belgium. pp 309.

Vanacker, V. (2002). Geomorphic response to human-induced environmental change in tropical mountain areas. The Austro Ecuatoriano as a case study. Doctoral thesis, *Katholic University of Leuven*, pp. III-1-VII-9pp.

Varnes, D. J. (1958). Landslides types and processes. In: *Landslide and Engineering Practice (edited by Eckel, F.B.)*. Highway Research Board, Special Report 29. *Washington DC*, pp. 20-27.

Varnes, D. J. (1978). Slope movement types and processes. In: *Landslide analysis and control (edited by Schuster, R.L. and Krizer, R.J.)*. Transportation Research Board, Special Report 176, national Academy of Sciences, *Washington DC*, pp 11-33.

Varnes, D. J. (1978). Slope movement typesand processes. In: *Landslide analysis and control (edited by Schuster, R. L. and Krizer, R .J.)*. Transportation Research Board, Special Report 176, national Academy of Sciences, *Washington DC*, pp. 11-33.

Verbesselt, J. & Mertens, K. (2000). *Towards land evaluation in the Uluguru Mountains Tanzania*. MSc dissertation, *Catholic University of Leuven, Belgium*, pp 116.

Westerberg, L. O. (1999). *Mass movements in East Africa highlands: Processes, effects and scar recovery*. Doctoral dissertation No. 14. The Department of physical Geography, *Stockholm University*, *Sweden*, pp 147.

Westerberg, L. O. & Christiansson, C. (1999). Highlands in East Africa: Unstable slopes, Unstable environments? *Ambio Vol.*, *28* No. *5*, 419-429.

Zinck, J. A. (1990). Soil survey: Epistemology of a vital discipline. *ITC Journal*, 1990-4, 335-351.

In: No-Till Farming
Editor: Earl T. Nardali

ISBN: 978-1-60741-402-5

Chapter 8

EFFECTS OF NO-TILL FARMING ON GLOBAL WEED-RELATED PROBLEMS AND WEED COMMUNITIES OF GREECE AND THE ONGOING CHALLENGES OF INTEGRATED WEED AND CROP MANAGEMENT

I. S. Travlos** and G. Economou

Laboratory of Agronomy, Department of Crop Production, Agricultural University of Athens, 75, Iera Odos st., 11855 Athens, Greece.

ABSTRACT

The changes of the cultivation methods and primarily the domination of no-till farming in many crops seem to dramatically influence, among others, the weed populations, the relative balance and frequency of the several weed species, and consequently the weed management practices which should be adopted in order to handle each new problem. This is the case in many Mediterranean climate regions, where no-tillage agriculture progressively becomes more and more popular and assigns the future evolutions. The example of the several species of *Conyza* (*C. albida*, *C. canadensis* and *C. bonariensis*), which have already taken advantages of the no-till farming practices and consecutively cover larger areas in Greece, is typical of this developing situation. Focusing on such a case study and particularly on the several aspects of the causes, consequences, distribution and management strategies for the above-mentioned and many other weed species, could definitely offer an opportunity for understanding the overall situation and seem crucial in order to think about and invoke several constituents of integrated weed management programs on a local and global basis. The subject of this study is to review the literature on some outstanding cases of weed species and problems that have arisen due to the no-till agriculture and to concisely present the example of the *Conyza* species.

---

* Corresponding Author. Present address: Benaki Phytopathological Institute, Department of Weed Science, Laboratory of Integrated Weaed Management and Plant Growth Regulators, 8 St. Delta street, GR-145 61 Kifissia, Athens, GREECE.

INTRODUCTION

Producers in many areas of the world are changing their crop cultivation methods because of the advantages of no-till production systems. No-till systems are especially favourable in some cases, since herbicides replace tillage, which leaves more crop residue on the soil surface. Crop residue increases both precipitation storage in soil and water use efficiency of crops (Peterson et al. 1996); consequently, producers have adopted similar practices in many crop fields, orchards and vineyards worldwide. Despite the several benefits of no-till systems, some considerations should be taken into serious account, since the managerial requirements in such a production system are generally greater than that of a conventional till system, especially in the case of weeds.

Indeed, the changes of the cultivation methods and primarily the domination of no-till farming for many crops seem to dramatically influence, among others, the weed populations, the relative balance and frequency of the several weed species, and consequently the weed management practices which should be adopted in order to handle each new problem. Under this view, which is the case in many Mediterranean climate regions too, whereas no-tillage agriculture progressively becomes more and more popular, the objective of this paper is to highlight some of the most interesting cases of recent and rapidly increasing weed problems which are relative to no- or minimum-till farming.

NO-TILL FARMING

No-till is lately a common practice for many crops and countries, since in many cases, crop production is increasing under no-till in response to integrated crop management techniques that are known to have greater economic advantages and better soil and moisture conserving properties than conventional tillage (Shear 1985). Besides, reduced and no-till farming systems minimize soil erosion on highly erodible farmland (Renard et al., 1994) and sediment contamination of surface water by reducing both primary tillage operations and field cultivation for weed management (Donald et al., 2001). Mulch can be compared to an umbrella that protects the soil from damage by the sun and the impact of the rain. Many farmers consider the long term impacts of their actions on the soil and environment. It is therefore highly recommended that farmers follow the pattern of nature and leave the crop residues on the land to decompose and restore the fertility of the soil. The crucial precautions which must be followed are related with the quantity and the size of crop residues, since the late decomposition could probably have adverse effects on crop productivity.

Cover crop residues, whether remaining on the soil surface or incorporated with tillage, also influence weed emergence. Fall-planted cover crops have been evaluated extensively for weed suppression in many systems and are commonly used to stabilize soil and capture soil nitrates during the fall and winter. Residues of fallow-season cover crops that remain on the soil surface reduce broadleaf weed emergence in summer crops by modifying the soil thermal and light environment (Teasdale and Mohler 1993) and by presenting a physical barrier to emerging weeds (Teasdale and Mohler 2000). Incorporating cover crop residues into the soil has been shown to both increase or decrease weed emergence depending on the amount and type of cover crop (Blum et al. 1997).

INVASIVE SPECIES AND WEEDS FAVORED BY NO-TILL PRACTICES—THE DANGER OF HERBICIDE RESISTANCE

Biological invasions are large-scale phenomena of widespread importance, which represent one of the major threats to European and global biodiversity. Numerous exotic plant species, many introduced into Europe little more than 200 years ago, have become successfully established over large areas of the European Community. Furthermore, the geographical range of a large number of species is believed to be increasing and some exotic species that currently have a limited distribution in Europe may, in the future, become more widespread. Regardless of the mechanism, the impact is clear of the invasive species on natural plant communities; they may also be an economic problem, becoming established as highly persistent and vigorous agricultural weeds, damaging manmade environments or choking open spaces and waterways. Several economic and environmental drivers markedly increase ecosystem vulnerability to invasion such as agriculture, urbanization, transport and climatic changes. The impact of all of these factors is increasing rapidly on a global scale. So far, many species cause major economic problems related to their prevention, control and eradication, such as damage to crops, damage in urban areas, congestion in waterways, etc. The species of *Conyza* (*C. canadensis*, *C. albida* and *C. bonariensis)* are among these weeds (Economou et al., 2008).

Soil tillage usually results in an increase in emergence of annual broadleaf weeds (Froud-Williams et al. 1984; Shilling et al. 1986). This increase in weed emergence is often synchronized with crop emergence and is attributed to the strong periodicity of germination in many summer annual weeds (Stoller and Wax 1973; Roberts and Bodrell 1983) coupled with seed exposure to germination promoters. Spring tillage increases the proportion of seeds that germinate and emerge by exposing weed seeds to germination promoters such as light (Scopel et al. 1994), increasing the probable zone of emergence for small-seeded weeds (Mohler and Galford 1997). If time is adequate in the spring, tillage can be used to increase weed emergence and reduce weed seedbanks before crops are planted (Mulugeta and Stoltenberg 1997). Elimination or reduction of soil tillage in the spring, as is common in no-till or other reduced tillage systems, should suppress annual broadleaf weed emergence by reducing the proportion of weed seeds that germinate and emerge. However, emergence of summer annual broadleaf weeds has been shown to both increase (Buhler 1992; Teasdale et al. 1991) and decrease (Froud-Williams et al. 1984; Mohler and Calloway 1992) when tillage is reduced. Accumulation of weed seeds with time near the soil surface in systems without soil tillage may partially explain these inconsistencies. Tillage systems that concentrate weed seeds near the soil surface, as typically occurs in no-till, may exhibit relatively high weed emergence compared with systems that do not accumulate seeds at the surface, even though the proportion of seeds emerging from the seedbank may be less in no-till than in conventional tillage (Mohler 1993).

The cause of reduced weed emergence for some species in no-till farming is likely due to the lack of germination stimulants that reduces the fraction of seed that germinates. Other causes could be soil crusting, a restricted emergence zone (also related to crusting), greater soil resistance to emergence, and slightly lower survival of weed seeds near the soil surface.

Conyza spp. (horseweed, Canadian fleabane, marestail, etc.) includes annual plants of the Asteraceae family and are often considered as winter annual weeds (Gibson et al. 2005).

However, Regeher and Bazzaz (1979) indicated horseweed germination can also occur in spring. Indeed, *Conyza* spp. can germinate under a wide variety of conditions (Main et al. 2006) and can complete their lifecycle as winter or summer annual species (Buhler and Owen 1997; Davis and Johnson 2008). It has long been a common species along roadsides and field edges, while horseweed seed is wind-disseminated and an individual plant is capable of producing more than 200,000 seeds (Bhowmik and Bekech 1993; Weaver 2001). Seeds germinate best in early fall or spring; however germination can occur throughout the year (Buhler and Owen 1997; Holm et al. 1997; Regehr and Bazzaz 1979). Soil tillage for establishment of summer annual crops disrupts the life cycle of winter annuals, such as horseweed, and research has demonstrated that fall or spring tillage controls emerged horseweed and other winter annual weeds (Brown and Whitwell 1988). No-tillage farming practices create an environment where soil is not disturbed and herbicides replace tillage for weed control before seeding an annual crop (Bazzaz, 1990). Winter annual plants that germinate during fall and survive the winter have a competitive advantage over spring germinating plants for space, water, light, and nutrients thereby suppressing the growth of summer annual crops.

Regeher and Bazzaz (1979) consider horseweed as a successional winter annual that can rapidly infest abandoned fields. Horseweed has become one of the most common and problematic weeds in agronomic crops (Weaver 2001) since it is adapted to periodically plant-free, undisturbed soil and becomes established with an absence of tillage in crop production (Brown and Whitwell 1988; Buhler 1992; Holm et al., 1997). Therefore, the opportunistic nature of horseweed in undisturbed areas makes it well-suited for becoming established in agricultural fields and surrounding areas, particularly in no-tillage crop production and other systems using conservation tillage (Bhowmik and Bekech 1993; Brown and Whitwell 1988; Bruce and Kells 1990; Weaver 2001; Barnes et al. 2004), while this weed seems susceptible to common tillage practices of conventional-tillage cropping systems (Kapusta 1979; Brown and Whitwell 1988).

Over the last two decades, the percentage of crop hectares using conservation tillage systems has significantly increased (CTIC 2004; Young 2006) making weeds like *Conyza* spp. of great concern. Today, horseweed has been documented in more than 70 countries and 40 different crops around the world (Holm et al. 1997), and is listed as being a problematic weed in cotton (*Gossypium hirsutum*), grain sorghum (*Sorghum bicolor*), corn (*Zea mays*), soybean (*Glycine max*), orchards and vineyards (Brown and Whitwell 1988; Buhler and Owen 1997; Mueller et al. 2003; Vencill and Banks 1994; Wiese et al. 1995).

Historically, horseweed has been of little agronomic importance because, in large part, it was controlled by conventional tillage cropping practices and because of essentially complete control with the herbicides glyphosate, 2,4-D, and dicamba in no-tillage cropping practices (Brown and Whitwell 1988; Bruce and Kells 1990; Scott et al. 1998; Van Gessel 2001). Unfortunately, farmers chiefly rely on herbicides to manage weeds in reduced or no-tilled crops. A dramatic increase in glyphosate use was the most obvious change of weed management, associated with the adoption of glyphosate-resistant crops and no-till farming. Consequently, the diversity of herbicides used for weed management in these crops has significantly declined, resulting in many new cases of herbicide resistant weeds (Young, 2006). In fact, many growers have abandoned the principles of sound weed and herbicide-resistance management and instead of incorporating glyphosate into a resistance management strategy utilizing multiple herbicide sites of action and other methods; they rely exclusively upon glyphosate for weed control.

According to Green (2007) and for more than two decades, there were no known cases of glyphosate-resistant weeds and their evolution did not seem likely (Bradshaw et al. 1997). No higher plants were identified with any significant metabolic capability to inactivate glyphosate (Dyer 1994). Additionally, the use pattern for glyphosate did not favor the evolution of glyphosate-resistant weeds (Bradshaw et al. 1997). Glyphosate was initially expensive and primarily targeted at perennial weeds where seed production and other adaptive strategies were slower than annuals or used at low rates in weed management programs with herbicides that had other modes of action. However, the situation changed in about 1996 with the discovery of glyphosate-resistant rigid ryegrass (*Lolium rigidum*) in Australia (Powles et al. 1998; Pratley et al. 1999). This resistance took fifteen years to evolve in response to two annual applications of glyphosate. In 1997, glyphosate-resistant goosegrass (*Eleusine indica*) was identified. Goosegrass took ten years to evolve glyphosate resistance with up to eight applications of glyphosate each year (Lee and Ngim 2000). To date, more than ten glyphosate-resistant weed species have been reported, with horseweed being the first glyphosate-resistant weed identified in a glyphosate resistant crop system (VanGessel 2001). Glyphosate-resistant horseweed populations appeared just 3 yr after the initial use of glyphosate-resistant soybeans, but the field had previously received glyphosate applications in another use pattern and cropping system. The single-gene, partially dominant nature of glyphosate-resistant horseweed (Zelaya et al. 2004) and achenes allowed rapid widespread dissemination by the wind and quick establishment of resistant populations on over two million hectares across the midwestern, southern, and Atlantic states in just five years after the first report (Heap 2006; VanGessel 2001).

Consequently, the reliance on glyphosate as the sole herbicide in glyphosate resistant crops has lead to the selection of horseweed with resistance to glyphosate in USA and other countries (Van Gessel 2001; Mueller et al. 2003; Koger et al. 2004; Main et al. 2004; Heap 2007). Thus, increasing adoption of herbicide-resistant crop technology, especially glyphosate-resistant crops, has encouraged reduced and no-till production practices. This switch has progressively encouraged traditionally noncropland weeds such as horseweed to move into and infest cropland. Compounding this problem, horseweed populations have developed resistance to glyphosate in several countries (Koger et al. 2004), being the first annual broadleaf documented as resistant to glyphosate (VanGessel 2001), and one of the top 10 worst weeds with regard to its ability to develop resistance to many different types of herbicide site of action (Heap 2007). Herbicide resistance in horseweed was first observed for bipyridyliums in 1980. *Conyza* spp. resistance to photosystem II inhibitors, ureas and amides, ALS inhibitors, and glycines has since occurred. Horseweed resistance to ALS inhibitors was observed in Israel in 1993 and Poland in 2000, but was not observed in the United States until 2001 and many other countries since then. Populations of horseweed are also resistant to paraquat (Lehoczki et al. 1992; Smisek et al. 1998) and triazines (Lehoczki et al. 1984).

Conversely, and among the several negative impacts of genetically modified crops, the use of herbicide-resistant crops could increase grower dependence upon chemical weed control and result in an herbicide-crop monoculture that increases dramatically the risk of developing herbicide-resistant weeds (Radosevich et al. 1992). Over the next few years it is likely that glyphosate use will continue to increase with the growing adoption of glyphosate-resistant crops as well as increases in glyphosate use rates. A continuation of the trend toward delaying glyphosate application will likely prompt a shift to weeds species that are more tolerant to glyphosate. Furthermore, the selection pressure imposed by extensive and exclusive glyphosate use will undoubtedly result in an increasing frequency of reports of glyphosate-resistant weed biotypes all over the world (Hilgenfeld et al., 2004). It is also noticeable that among the world's greatest herbicide resistance problem is in Australia, where there is the world's highest adoption of no-tillage, too. Unfortunately, despite no-till has been essential to conserve fragile soils and for a number of other agronomic advantages (such as early seeding), it removes cultivation for weed control and no-till systems make for higher reliance on herbicides.

STRATEGIES, TRENDS AND CHALLENGES OF INTEGRATED WEED & CROP MANAGEMENT

Weed resistance affects thousands of fields and the most widely used herbicide products (Heap 2006). Herbicide-resistant weeds can increase the cost of crop production, limit the types of crops that can be grown, lower yields, and even lower land values (Anonymous 2002). Today, nobody really knows how widespread herbicide-resistant weeds are, but most weed scientists agree that herbicide resistant weed populations are progressively becoming more common (Green, 2007). Herbicide-resistant weeds are an immediate economic problem, especially when other desirable weed management options are not accessible (e.g. tillage).

Talking about weed management in such cases, there is certainly no one-size-fits-all program. It should be taken into account that weed ecology under no-till differs considerably from conventionall tillage (Staniforth and Wiese 1985). Weed emergence under no-till continues over a several weeks, making it difficult to decide on the best time to apply herbicides to achieve optimum weed control. For example, a tank-mix of glyphosate with residual herbicides may be needed to control late-emerging weeds that often become major problems (VanGessel et al., 2001). Another key component is that all weed management programs need to change over time as weeds adapt. Generally, best management practices encourage the grower to use diverse control methods in an economic and sustainable manner, and these practices can be referred to as integrated weed management (IWM). Therefore, the best strategy seems to be the seasonable management of the evolution of herbicide resistance, which means prevention and not waiting until the problem is recognized. Such strategy focuses on managing herbicide selection pressure (Green, 2007).

Weed resistance is a problem when a single herbicide mode of action is used repeatedly over several years. Any weed control method that reduces the selection pressure from that mode of action will reduce the rate of herbicide-resistance evolution. One of the most common and economical methods to manage herbicide-resistant weeds is often to tank-mix a herbicide to improve control (Culpepper 2006). By adopting the use of herbicide mixtures with different modes of action, selection pressure for any specific herbicide mode of action is reduced and herbicide utility may be preserved (Diggle et al. 2003).

The use of any type of cover crop or organic plant residue has also been suggested. In each case, it is necessary to test for the right amount and application. This method is appropriate for use in vegetables and trees. The covers of harvest residues and other organic materials affect the germination, the survival, growth and the competitive ability of weeds.

Generally, it seems that the toxic effect of the residues is higher against weeds with small seeds than those with big seeds, while it seems that the management of harvest residues offers a good opportunity for weed suppression (Liebman and Mohler, 2001). The effect of these covers over the weeds can be classified as direct and indirect.

The direct effects are those produced by the interaction of the covers with the weeds or their seeds, while indirect effects are caused by environmental modification where weeds grow due to the presence of the cover in the soil. In addition, the use of harvest residues as soil cover plays an important role in soil conservation. Talking about cover crops, in Greece, *Oxalis pes-caprae* is one of the most dominant weeds in the olive grooves and vineyards (Damanakis and Markaki, 1990), while it often forms dense mats in any area with open space. This high competitive ability seems of great potential and has to be investigated and further exploited, as long as *O. pes-caprae* has detrimental impacts by causing soil enrichment and stabilization of semi-stable areas and orchards and suppressing other weeds (Paspatis, 1987; Travlos et al., 2008).

Despite the plenty alternative methods for the single use of herbicides, none of them is enough for an absolute solution; their application requires integration into a diversified strategy (Economou et al., 2002; Economou et al., 2007; Travlos et al., 2007; Travlos and Paspatis, 2008; Travlos et al., 2009). The technical agent should choose the method that is the most appropriate and adaptable to its environment, taking into consideration economical criteria and assessing possible advantages and disadvantages. These methods may reduce the need for herbicidal use and help to reduce the selection pressure on the weeds and possible cases of resistance. Obviously, when these methods are applied, new problems may emerge. Therefore and in order to prevent resistance, weed management activities should be planned for the medium and long term, but not annually, integrating various weed control strategies. Strategies should be applied as diversely as possible in order to minimize the risk of resistance.

It is important to take into account that the repetition of any control method, be it mechanical, cultural or chemical, usually results in the adaptation of a weed species. The best tool for management is the frequent shift of strategy. Integrated weed management (IWM) is considered the most effective way to control weeds, which is based on the coordinated use of different strategies in the medium and long term, including a combination of various methods, direct and indirect, to keep weed infestation below the admissible economic threshold (Zwerger, 1996). In all these cases, a maximum of possible control methods should be combined, which may accordingly provide sustainable weed control in time. A program of integrated management should be:

- flexible, able to respond to the crop requirements at the time of its application;
- based on a good understanding of the biology of the target weed species;
- adapted to the farm conditions: climate, soil, history;
- linked to the economical goals of long-term farming;
- inexpensive (with a good cost-benefit ratio) in the short term (Fernández-Quintanilla et al., 1999; CRC, 2006).

Although specific recommendations vary for different weeds and for different regions and crops, according to Green (2007) most recommendations are relative to:

1 Start with a clean field and control weeds early.
2. Follow label directions. Apply herbicides at the proper rates and application timings, i.e., the right rate at the right time.
3. Do not rely on a single herbicide mode of action. Use mixtures or sequential applications of herbicides with different modes of action, soil residual, and overlapping weed spectrum.
4. Use crop rotations and cover crops when possible. Different crops allow different cultural and tillage options and some compete aggressively with weeds.
5. Control weeds throughout the season to reduce the seedbank. Control weed escapes before they can set seeds.
6. Use clean seeds and clean equipment to minimize the spread of weed seed.
7. Keep good records so any changes in weed patterns can be quickly recognized and managed.

According to the new challenges, several questions have arisen. One of them has to do with ideal level of required weed control (Green, 2007). Historically, weed management models often considered weed control for only the current season and recommended using only the lowest rate of herbicide needed to control weeds at economic thresholds, i.e. the lowest rate needed to protect yields (Buhler 2002). This approach used the surviving herbicide-sensitive weeds and herbicide-sensitive seeds in the seedbank to dilute the impact of herbicide-resistant weeds. However, the surviving herbicide-resistant seeds create a long-term management problem (Norris 1999). Additionally, growers need a very high degree of biological understanding to control weeds at the economic threshold (Coble 1994). New models are addressing the issue of herbicide-resistant weeds and seed production from a multiyear perspective, with more emphasis on managing weed seed production (Brainard and Bellinder 2004; Doyle and Stypa 2004).

Research where weed seed production was accurately determined suggests that several important weed species should be managed for zero weed seed production (Norris 1999). Application of economic thresholds to an invading weed species could be disastrous if it allows a difficult-to-control weed to survive and establish a seed bank before a grower takes any action to control. According to the new trends of world agriculture (Green, 2007), for an invading herbicide-resistant weed biotype that is expanding its range, the use of zero seed threshold may always be appropriate in order to prevent the spread of weed resistance.

In conclusion and under the ongoing expanding of no-till systems, weed management is a perennial challenge for growers, and continual innovation is essential to maintain the effectiveness of management technologies. It is clear that a farming system with diversity can be robust. If enough different tools are used, weed resistance and other problems can be kept at relatively low levels. In many cases, herbicides remain the main tool for fighting weeds but the key is rotation of herbicide groups and other alternative methods. With any weed-control tool, especially herbicides, diversity is the key word and the objective is to continually confuse the weeds by diversifying the weed control strategies.

References

Anonymous. (2002). Managing Glyphosate-Resistant Weeds-An Investment in Land Value. Syngenta Crop Protection. www.syngentacropprotection-us.com/ Resources/Prod/ Touchdown/LandValues.pdf. Accessed: October 31, 2005.

Barnes, J., Johnson, B., Gibson, K. & Weller, S. (2004). Crop rotation and tillage system influence late-season incidence of giant ragweed and horseweed in Indiana soybean. *Crop Management,* doi, 10.1094/CM-2004-0923-1002-BR.

Bazzaz, F. A. (1990). Plant-plant interaction in successional environments. In: Grace, J. B. & Tilman, D. (Eds.). *Perspectives on Plant Competition*, pp. 239-263, San Diego, CA: Academic.

Bhowmik, P. C. & Bekech, M. M. (1993). Horseweed (*Conyza canadensis*) seed production, emergence and distribution in no-till and conventional-tillage corn (*Zea mays*). *Agronomy Trends Agricultural Science*, *1*, 67-71.

Blum, U., King, D. L., Gerig, T. M., Lehman, M. E. & Worsham, A. D. (1997). Effects of clover and small grain cover crops and tillage techniques on seedling emergence of some dicotyledonous weed species. *American Journal of Alternative Agriculture*, *12*, 146-161.

Bradshaw, L. D., Padgette, S. R., Kimbal, S. L. & Wells, B. H. (1997). Perspectives on glyphosate resistance. *Weed Technology*, *11*, 189-198.

Brainard, D. C. & Bellinder, R. R. (2004). Assessing variability in fecundity of *Amaranthus powelli* using a simulation model. *Weed Research*, *44*, 203-217.

Brown, S. M. & Whitwell, T. (1988). Influence of tillage on horseweed (*Conyza canadensis*). *Weed Technology*, *2*, 269-270.

Bruce, J. A. & Kells, J. J. (1990). Horseweed (*Conyza canadensis*) control in notillage soybeans (*Glycine max*) with preplant and preemergence herbicides. *Weed Technology*, *4*, 642-647.

Buhler, D. D. (1992). Population dynamics and control of annual weeds in corn (*Zea mays*) as influenced by tillage systems. *Weed Science*, *40*, 241-248.

Buhler, D. D. (2002). Challenges and opportunities for integrated weed management. *Weed Science*, *50*, 273-280.

Buhler, D. D. & Owen, M. D. K. (1997). Emergence and survival of horseweed (*Conyza canadensis*). *Weed Science*, *45*, 98-101.

Coble, H. D. (1994). Future directions and needs for weed science research. *Weed Technology, 8*, 410-412.

CRC for Australian Weed Management. (2006). *Integrated weed management manual.* (available at http://www.weeds.crc.org.au). Consulted on 27 December 2006.

CTIC (2004). 2004 soybeans—ranked by percentage of no-till acres. West Lafayette, IN: http://www.in.gov/isda/files/soybeanrankpercent.pdf. Accessed: September 24, 2007.

Culpepper, A. S. (2006). Glyphosate-induced weed shifts. *Weed Technology*, *20*, 277-281.

Damanakis, M. & Markaki, M. (1990). Studies on the biology of *Oxalis pes-caprae* L., under field conditions in Crete. *Zizaniology*, *2*, 145-154.

Davis, V. M. & Johnson, W. G. (2008). Glyphosate-resistant horseweed (*Conyza canadensis*) emergence, survival, and fecundity in no-till soybean. *Weed Science*, *56*, 231-236.

Diggle, A. J., Neve, P. B. & Smith, E. P. (2003). Herbicides used in combination can reduce the probability of herbicide resistance in finite weed populations. *Weed Research*, *43*, 371-382.

Donald, W. W., Kitchen, N. R. & Sudduth, K. A. (2001). Between-row mowing and banded herbicide to control annual weeds and reduce herbicide use in no-till soybean (*Glycine max*) and corn (*Zea mays*). *Weed Technology*, *15*, 576-584.

Doyle, P. & Stypa, M. (2004). Reduced herbicide rates - a Canadian perspective. *Weed Technology*, *18*, 1157-1165.

Dyer, W. E. (1994). Resistance to glyphosate. In: Powles, S. B. & Holtum, J. A. M. (Eds.), *Herbicide Resistance in Plants*, pp. 229-241. Boca Raton, FL: Lewis.

Economou, G., Tzakou, O, Gani, A., Yannitsaros, A. & Bilalis, D. (2002). Allelopathic effect of *Conyza albida* on *Avena sativa* and *Spirodela polyrhiza*. *Journal of Agronomy & Crop Science*, *188*, 248-253.

Economou, G., Travlos, I. S., Folinas, A. & Karamanos, A. I. (2007). Greek Oregano (*Origanum vulgare ssp. hirtum*) as an allelopathic plant. *Journal of Food, Agriculture & Environment*, *5(1)*, 84-87.

Economou, F., Travlos, I. S. & Kotoula, E. (2008). The current situation of invasive species in Greece. *2nd International Symposium of Intractable Weeds and Plant Invaders, European Weed Research Society*, Osijek, Croatia, September 14 to 18, 2008.

Fernández-Quintanilla, C., Garrido, M. & Zaragoza, C. (1999). *Control integrado de las malezas. Buenas Prácticas Agrícolas*. Phytoma España S.L. *Valencia, Spain*, 271 pp.

Froud-Williams, R. J., Chancellor, R. J. & Drenan. D. S. H. (1984). The effects of seed burial and soil disturbance on emergence and survival of arable summer annual weeds in relation to minimal cultivation. *Journal of Applied Ecology*, *21*, 629-641.

Gibson, K. D., Johnson, W. G. & Hillger, D. E. (2005). Farmer perceptions of problematic corn and soybean weeds in Indiana. *Weed Technology*, *19*, 1065-1070.

Green, J. M. (2007). Review of glyphosate and ALS-inhibiting herbicide crop resistance and resistant weed management. *Weed Technology*, *21*, 547-558.

Heap, I. M. (2006). The international survey of herbicide resistant weeds. http://www.weedscience.com. Accessed: September 15, 2006.

Heap, I. (2007). Herbicide Resistant Weeds. Weed Science Society of America: Web Page: http://www.weedscience.org/. Accessed: March 5, 2007.

Hilgenfeld, K. L., Martin, A. R., Mortensen, D. A. & Mason, S. C. (2004). Weed management in a glyphosate-resistant soybean system: weed species shifts. *Weed Technology*, *18*, 284-291.

Holm, L., Doll, J., Holm, E., Pancho, J. & Herberger, J. (1997). *World Weeds: Natural Histories and Distributions*. New York: *J. Wiley*, 1129 p.

Kapusta, G. (1979). Seedbed tillage and herbicide influence on soybean (*Glycine max*) weed control and yield. *Weed Sciencei 27*, 520-526.

Koger, C. H., Poston, D. H., Hayes, R. M. & Montgomery, R. F. (2004). Glyphosate-resistant horseweed (*Conyza canadensis*) in Mississippi. *Weed Technology*, *18*, 820-825.

Lee, L. J. & Ngim, J. (2000). A first report of glyphosate-resistant goosegrass (*Eleusine indica* (L.) Gaertn) in Malaysia. *Pest Management Science*, *56*, 336-339.

Lehoczki, E., Laskay, G., Polos, E. & Mikulas, J. (1984). Resistance to triazine herbicides in horseweed (*Conyza canadensis*). *Weed Science, 32*, 669-674.

Lehoczki, E., Laskay, G., Gaal, I. & Szigeti, Z. (1992). Mode of action of paraquat in leaves of paraquat-resistant *Conyza canadensis*. *Plant Cell Environment*, *15*, 531-539.

Liebman, M. & Mohler, C. (2001). Weeds and the soil environment. In: M. Liebman., C. & Mohler, C. Staver (Eds.), *Ecological management of agricultural weeds*. Cambridge, *United Kingdom*, Cambridge University Press.

Main, C. L., Mueller, T. C., Hayes, R. M. & Wilkerson, J. B. (2004). Response of selected horseweed (*Conyza canadensis* (L.) Cronq.) populations to glyphosate. *Journal of Agriculture & Food Chemistry*, *52*, 879-883.

Mohler, C. L. (1993). A model of the effects of tillage on emergence of weed seedlings. *Ecological Applications*, *3*, 53-73.

Mohler, C. L. & Calloway, M. B. (1992). Effects of tillage and mulch on the emergence and survival of weeds in sweet corn. *Journal of Applied Ecology*, *29*, 21-34.

Mohler, C. L. & Galford, A. E. (1997). Weed seedling emergence and seed survival: separating the effects of seed position and soil modification by tillage. *Weed Research*, *37*, 147-155.

Mueller, T. C., Massey, J. H., Hayes, R. M., Main, C. L. & Stewart, C. N. (2003). Shikimate accumulates in both glyphosate-sensitive and glyphosate-resistant horseweed (*Conyza canadensis* L. Cronq.). *Journal of Agriculture & Food Chemistry*, *51*, 680-684.

Mulugeta, D. & Stoltenberg, D. E. (1997). Increased weed emergence and seed bank depletion by soil disturbance in no-tillage systems. *Weed Science*, *45*, 234-241.

Norris, R. F. (1999). Ecological implications of using thresholds for weed management. *Journal of Crop Production*, *2*, 31-58.

Paspatis, E. A. (1987). Chemical, cultural and biological control of *Oxalis pes-caprae* in vineyards in Greece. In: Weed Control on Vine and Soft Fruits, *Proceedings of a Meeting 10 of the EC Experts' Group, Commision of the European Communities, Dublin*, 12-14 June 11 1985, pp. 27-29.

Peterson, G. A., Schegel, A. J., Tanaka, D. L. & Jones, O. R. (1996). Precipitation use efficiency as affected by cropping and tillage systems. *Journal of Production Agriculture*, *9*, 180-186.

Powles, S. B., Lorraine-Colwill, D. F., Dellow, J. J. & Preston, C. (1998). Evolved resistance to glyphosate in rigid ryegrass (*Lolium rigidum*) in Australia. *Weed Science*, *46*, 604-607.

Pratley, J. E., Urwin, N. A. R., Stanton, R. A., Baines, P. R., Broster, J. C., Cullis, K., Schafer, D. E., Bohn, J. A. & Krueger, R. W. (1999). Resistance to glyphosate in Lolium rigidum: I. Bioevaluation. *Weed Science*, *47*, 405-411.

Radosevich, S. R., Ghersa, C. M. & Comstock, G. (1992). Concerns a weed scientist might have about herbicide tolerant crops. *Weed Technology*, *6*, 635-639.

Regehr, D. L. & Bazzazz, F. A. (1979). The population dynamics of *Erigeron canadensis*, a successional winter annual. *Journal of Ecology*, *67*, 923-933.

Renard, K. G., Laflen, J. M. Foster, G. R. & McCool, D. K. (1994). The revised universal soil loss equation. In: R. Lal, (Ed.). *Soil Erosion Research Methods*. 2nd ed. pp. 105-124. St. Lucie Press, Delray Beach, FL.

Roberts, H. A. & Bodrell, J. E. (1983). Field emergence and temperature requirements for germination in *Solanum sarrachoides. Sendt. Weed Research*, *23*, 247-252.

Scopel, A. L., Ballare, C. L. & Radosevich, S. R. (1994). Photostimulation of seed germination during soil tillage. *New Phytologist*, *126*, 145-152.

Shear, G. M. (1985). Introduction and history of limited tillage. In: A.F. Wiese, (Ed.). *Weed Control in Limited-Tillage Systems*, pp. 1-14. Champaign, IL: Weed Science Society of America.

Shilling, D. G., Worship, A. D. & Danehower, D. A. (1986). Influence of mulch, tillage and diphenamid on weed control, yield, and quality in no-till flue cured tobacco (Nicotiana tabacum). *Weed Science*, *34*, 738-744.

Smisek, A., Doucet, C., Jones, M. & Weaver, S. (1998). Paraquat resistance in horseweed (*Conyza canadensis*) and Virginia pepperweed (*Lepidium virginicum*) from Essex County, Ontario. *Weed Science*, *46*, 200-204.

Staniforth, D. W. & Wiese, A. F. (1985). Weed biology and its relationship to weed control in limited-tillage systems. In: A.F. Wiese (Ed.) *Weed Control in Limited-Tillage Systems*, pp. 15-25. Champaign, IL: Weed Science Society of America.

Stoller, E. W. & Wax, L. M. (1973). Periodicity of germination and emergence of some annual weeds. *Weed Science, 21*, 574-580.

Teasdale, J. R., Best, C. E. & Potts, W. E. (1991). Response of weeds to tillage and cover crop residue. *Weed Science*, *39*, 195-199.

Teasdale, J. R. & Mohler, C. L. (1993). Light transmittance, soil temperature, and soil moisture under residue of hairy vetch and rye. *Agronomy Journal*, *85*, 673-680.

Teasdale, J. R. & Mohler, C. L. (2000). The quantitative relationship between weed emergence and the physical properties of mulches. *Weed Science*, *48*, 385-392.

Travlos, I. S., Economou, G., Kanatas P. J. & Tzakou, O. (2007). Aspects of the allelopathic potential of horseweed (*Conyza albida*). *International Journal of Agricultural Research*, *2(4)*, 397-401.

Travlos, I. S. & Paspatis, E. A. (2008). Aspects of the allelopathic effects of Heliotrope (*Heliotropium europaeum* L.) on *Avena sativa*, *Phaseolus vulgaris* and *Spirodela polyrhiza*. *Allelopathy Journal*, *21(2)*, 397-404.

Travlos, I. S., Paspatis, E. & Psomadeli, E. (2008). Allelopathic potential of *Oxalis pes-caprae* tissues and root exudates. *Journal of Agronomy*, *7*, 202-205.

Travlos, I. S., Economou, G., Kotoulas V. E., Kanatas, P. J. Kontogeorgos, A. N. & Karamanos, A. I. (2009). Potential effects of diurnally alternating temperatures and solarization on purple nutsedge (*Cyperus rotundus*) tuber sprouting. *Journal of Arid Environments*, *73*, 22-25.

VanGessel, M. J. (2001). Glyphosate-resistant horseweed from Delaware. *Weed Science*, *49*, 703-705.

Vencill, W. K. & Banks, P. A. (1994). Effects of tillage systems and weed management on weed populations in grain sorghum (*Sorghum bicolor). Weed Science*, *42*, 541-547.

Weaver, S. E. (2001). The biology of Canadian weeds. 115. *Conyza canadensis*. *Canadian Journal of Plant Science*, *81*, 867-875.

Wiese, A. F., Salisbury, C. D. & Bean, B. W. (1995). Downy brome (*Bromus tectorum*), jointed goatgrass (*Aegilops cylindrica*), and horseweed (*Conyza canadensis*) control in fallow. *Weed Technology*, *9*, 249-254.

Young, B. G. (2006). Changes in herbicide use patterns and production practices resulting from glyphosate-resistant crops. *Weed Technology*, *20*, 301-307.

Zelaya, I. A., Owen, M. D. K. & VanGessel, M. J. (2004). Inheritance of evolved glyphosate resistance in *Conyza canadensis* (L.) *TAG Theoretical and Applied Genetics*, *10*, 58-70.

Zwerger, P. (1996). Integrated weed management in developed nations. *Proc. 2nd International Weed Control Congress, Copenhagen,Denmark*, pp. 933-942.

In: No-Till Farming
Editor: Earl T. Nardali

ISBN: 978-1-60741-402-5

Chapter 9

EFFECTS OF AGROECOLOGICAL CONDITIONS AND SLOPE GRADIENT ON TILLAGE EROSION IN THE NORTHERN SLOPES OF THE ULUGURU MOUNTAINS, TANZANIA

***D. N. Kimaro*[a]*, J. Poesen*[b]*, J. A. Deckers*[c] *and H. B. Msita*[a]**

[a]Department of Agricultural Engineering and Land Planning, Sokoine University of Agriculture, P.O. Box 3003, Morogoro, Tanzania

[b]Physical and Regional Geography Research Group, Katholieke Universiteit Leuven, Geo-Institute, Celestijnenlaan 200 E, B-3001 Heverlee, Belgium

[c]Division of Soil and Water Management, Geo-Institute, Katholieke Universiteit Leuven, Celestijnenlaan 200 E, B-3001 Heverlee, Belgium

ABSTRACT

A study was conducted in the northern slopes of the Uluguru Mountains, Tanzania to investigate the effects of slope gradient, tillage depth, soil surface conditions and soil moisture on the rate of soil flux due to manual tillage. Soil flux was measured by means of an on-farm experiment in the farmers' fields. Twenty four plots (1.2 m wide and 8 m long) at different slopes: 31, 47, 51, 54, 58, 61, 65 and 67 percent were demarcated prior seasonal land preparation. Two sets of treatment at each slope site were done: deep tillage and shallow tillage under conditions of bare soil surface and soil surface with residues to both treatments. Effect of soil moisture condition was studied in areas under intensive agriculture at slopes 54 and 67 percent. Soil flux rates were monitored by Gerlarch trough method. Descriptive statistics (means, standard deviations, minimum and maximum) were widely employed in the exploratory analysis of soil flux rates and related factors. The effect of slope gradient and treatments were statistically analysed according to General Linear Model (GLM) of the Statistical Analysis System (SAS, 1999). Least Squares Means (LSMean) was used to compare the treatment means at 5 percent level of probability. The results showed that rates of soil flux due to tillage erosion generally increased with increasing slope gradient. Mean soil flux increased significantly ($P<0.05$) from 16 kg/m/tillage pass at slope gradient 31% to more than 60 kg/m/tillage pass at slope gradient >60%. The rates of soil flux from bare soil surfaces for both shallow and deep tillage were significantly higher ($P<0.05$) than soil flux from soil surface with residues. Bare soil surface promotes tillage erosion than soil surface with residues. Further analysis showed that a combination of shallow tillage and presence of residues on the soil surface can cut down the rates of soil flux by 60% compared to deep tillage with bare soil surfaces. Therefore, farmers should be discouraged to clear their fields by removing all the residues prior to cultivation. The study also revealed that soil flux increased with moisture content, a condition attributed to greater tillage depth.

Keywords: Tillage erosion, agro-ecological conditions, tillage, Uluguru Mountains, Tanzania

1.0. Introduction

1.1. General Overview on Tillage Erosion

Tillage erosion also referred to as "arable erosion" (Zachar, 1982) or mechanical soil erosion (Kiburys, 1995) is the process of soil movement caused by the force applied by agricultural tools and by gravity (Lindstrom *et al.*, 1992; Govers *et al.*, 1994, Dercon, 2001). At the level of farm plot, tillage erosion can be recognised by a less fertile tillage step at the top of the field (due to profile truncation) and the development of a soil bank with a high soil fertility status at the bottom of the field if the physical barrier is present (Turkelboom, 1999; Nyssen et al., 2000; Poesen *et al.*, 2000; Nyssen, 2001).

Although tillage erosion became better documented recently (Govers *et al.*, 1994; Guiresse and Revel, 1995; Govers *et al.*, 1996) there is still inadequate information on this process particularly in the tropical mountainous areas (Nyssen, 2001). The study by Govers *et al.* (1994) in Western Europe demonstrated that movement and redistribution of soil by tillage using tractor drawn implements can be described by diffusion-type equation as shown in equation (1) below.

$$Qt = k.S \tag{1}$$

Where Qt = Soil flux caused by tillage (kg/m/tillage pass)
K = Diffusion coefficient and
S = Slope gradient

The authors reported that because the soil flux is directly proportional to the slope gradient, erosion and deposition are also proportional to the slope gradient change. This means erosion occurs at the convexities, while deposition takes place at the concavities (Govers *et al.,* 1994). Also, zones with zero-influx such as hill tops and upper field boundaries face accelerated truncation of the upper soil horizons, while soil banks are commonly formed at lower field boundaries if a barrier is present (Guiresse and Revel, 1995). With respect to soil loss through tillage erosion, most studies to-date have concentrated on tractor-plough tillage on gentle slopes (Govers *et al.,* 1994; Guiresse and Revel, 1995; Poesen *et al.,* 1997). Only few works have dealt with soil loss caused by manual tillage (Turkelboom *et al.,* 1997; 1999; Turkelboom, 1999) which is presumably the most widespread method of land preparation in the tropical mountainous areas. Most tillage erosion studies have mainly focused on the effect of topography and cultivation practices on soil flux during tillage operations. Until recently, experimental studies on tillage erosion were mostly related to the design and performance of soil working implements for mechanised agriculture (Van Muysen *et al.*, 1999).

However, Turkelboom *et al.* (1999) attempted to quantify soil movement by manual hoeing as a function of slope angle and slope length. Working on a slope range of 17% to 82%, the authors observed that soil flux by manual deep tillage is quantified by linear functions for different slope gradient classes. This is opposed to the single diffusion-type of equation often proposed for tractor derived tillage erosion (Govers *et al.,* 1994) as shown on equation (2) below.

$$Qt = 77.0S + 13.3 \quad (\text{if } 3\% < S < 70\%) \tag{2}$$

Where: Qt = Soil flux caused by manual deep tillage (kg/m/tillage pass)
S = Slope gradient of original soil surface (%)

According to Nyssen *et al.* (2000a) tillage erosion by manual hoeing and/or oxen plough could be held responsible for half of the sediment deposited behind newly constructed stone bunds in the Ethiopian highlands.

As the first purpose of tillage is to control weeds, the pressure and type of weed will influence the tillage frequency and hence, tillage erosion rates (Dercon, 2001). Any measure that reduces the weed intensity, will automatically lead to reduced soil loss by tillage (Poesen *et al.*, 2000). Acording to Turkelboom *et al.* (1999), environmental friendly methods such as fallow cultivation systems, time of slushing, mulching and fertility management can control weeds and thus reduce the rate of tillage erosion. A switch to minmum or no-tillage cultivation, application of herbicides and perennial crop cultivation system are also reported as potential solutions (Tsara *et al.*, 2001). Other research work have reported the importance of using buffer strip system, grass barrier along contour and stone bunds on tillage erosion (Nyssen *et al.,* 2000; Poesen *et al.*, 2000; Nyssen, 2001). A peculiar phenomenon that has been observed in these soil conservation structures is the formation of progressive terraces (Turkelboom *et al.,* 1999; Nyssen *et al.,* 2000). Progressive terrace formation is commonly explained by sediment trapping and accumulation along contour bunds with grass barriers or hedgerows (Poesen *et al.*, 2000). Although trapping of water eroded sediment has been observed in the field, it is believed that this process is not the main cause for progressive terrace formation but tillage erosion can contribute significantly to the amount of sediment trapped behind these soil conservation structures (Dercon, 2001; Nyssen, 2001; Kimaro *et al.,* 2005).

1.2. Factors Influencing Tillage Erosion

Investigations conducted on gentle slopes and for mechanised agriculture indicated that tillage erosion is influenced by a number of factors including slope gradient (Govers *et al.,*

1994; Kachanoski *et al.* 1992), slope length (Poesen *et al.*, 2000), cultivation practices (Turkelboom *et al.*, 1999) and initial soil conditions (Van Muysen *et al.*, 1999). Kachanoski *et al.* (1992) while studying the effect of tillage implement on soil flux in south-western Ontario, Canada, observed that there was some indication that soil flux increased as slope gradient increased. The effect of slope gradient on soil translocation by tillage has also been observed under various agri-environmental conditions in North America (Lindstrom *et al.*, 1990), Western Europe (Govers *et al.*, 1994; Guiresse and Revel, 1995) and the Mediterranean (Poesen *et al.*, 1997).

In the tropics only few studies have been carried out (Nyssen *et al.,* 2000; Turkelboom *et al.*, 1999) relating soil flux and slope gradient. A study in the Ethiopian highlands on soil flux rates due to tillage by oxen ploughing indicated soil flux rates ranging from 4.8 kg/m on a 3% slope to 38.7 kg/m on a 48% slope (Nyssen *et al.,* 2000). Studying tillage erosion by manual hoeing in northern Thailand for the slope gradient class between 3% and 70%, Turkelboom *et al.* (1999) showed that soil flux rates increased from 22 kg/m/tillage pass at a slope of 31% to 78 kg/m/tillage pass at a slope of 71%.

Different tillage erosion rates have been reported from different cropping systems in Thailand (Turkelboom *et al.*, 1999). Working in northern Thailand, Poesen *et al.* (2000) showed that while using manual hoeing cultivation of upland rice and cabbage caused the highest soil transport followed by ginger and maize while the cultivation of beans resulted in the least soil flux. The authors argue that the observed differences could be attributed to different soil management practices associated with each cropping system. Van Muysen *et al.* (1999) working in Portugal observed that soil flux was 3 to 4 times higher when tillage is done on a freshly tilled, loose soil than on soil under grass fallow.

Studies of tillage erosion in the Sub Saharan Africa are scanty. The same is true with Tanzania. Though the contribution of mountainous areas to agriculture in Tanzania is enormous these are the areas where tillage erosion is expected to be very active. There is a wide range of studies on tillage erosion that ought to be carried out prior to adopting the appropriate tillage erosion control approaches. For example studies on the possible effects of tillage depth, surface and soil conditions on soil flux during manual tillage particularly in mountainous areas are of paramount importance for developing strategic soil erosion control measures associated with tillage operations.

Manual tillage is a wide spread method of land preparation in Tanzania, notably in mountainous areas. Little quantitative information is currently available on agro-ecological conditions influencing soil translocation by manual tillage. In the Uluguru Mountains agriculture is practised on very steep slopes up to 80%. Land preparation in these areas is done manually using hand hoes during dry season when weed pressure is low. Some farmers remove residues and/or set their fields on fire prior to cultivation. There is therefore a need to assess the effects of different agro-ecological conditions influencing tillage erosion in the Uluguru Mountains as a contribution to the overall development of soil conservation strategies in the area and others with similar characteristics.

This paper reports the results of tillage experiment which was set up to investigate the effects of slope gradient, tillage depth, surface ground conditions and soil moisture on tillage erosion due to manual hoeing in the Uluguru Mountains.

2.0. The Study Area

The study area is located on the northern slopes of the Uluguru Mountains between 350295E and 354368E and 9237500N and 9243697N UTM coordinates (Figure 1). The climate of the area is categorised as sub-humid tropical savannah of the low latitude environment (Sharma, 1987). The mean annual rainfall varies with altitude, from 900 mm at around 550 m asl to 2,300 mm at 1,500 m asl. It is distributed into 2 distinct periods, a long rain season (*masika*) which lasts from March to May and short rains (*vuli*) which extend from October to January (Figure 2). The mean annual temperature varies from 25°C around 550 m asl to 19°C at 1,500 m asl. The underlying bedrocks are metasediments consisting mainly of hornblende pyroxene granulites, with plagioclase and quartz rich veins (Sampson and Wright, 1964). The area is mountainous comprising strongly dissected mountain ridges and foothills with very steep narrow valleys.

Based on the World Reference Base for Soil Resources system of soil classification (FAO *et al.,* 1998), the soils on the mountain ridges are dominantly *Endoskeletic* and *Leptic Cambisols*. But there are also minor soil units *Haplic* and *Chromic Phaeozems* and *Orthieutric Regosols*. On the foothills the dominant soils are *Chromic Lixisols* and *Profondic Acrisols* associated with *Hyperferralic Cambisols* and *Endoleptic Cambisols* (Kimaro *et al.*, 1999).

To a great extent the Uluguru Mountains are under cultivation. The mountain ridges are mainly used for production of vegetables, beans, short rains maize and banana while on the foothills long rains maize is the main crop (Kimaro *et al.*, 1999).

3.0. Tillage Experiment

Soil flux by manual tillage was measured in the northern slopes of the Uluguru Mountains by means of an on-farm experiment in the farmers' fields during the dry season of the year 2000 and 2001. Twenty four plots (1.2 m wide and 8 m long) at different slopes (31, 47, 51, 54, 58, 61, 65 and 67 %) were demarcated prior to seasonal land preparation. Two sets of treatment at each slope site were done: deep tillage and shallow tillage under conditions of bare soil surface and soil surface with residues to both treatments. In the field, local farmers were requested to till the plots in a traditional way. The plots were tilled starting from the bottom of the field and moving up the slope. Farmers used a hoe with a steel blade of about 18.5 cm long and 16.5 cm wide and a wooden handle about 100 cm long. In order to measure soil translocation, the Gerlarch trough method (Gerlarch, 1967) was used. This method provides quick assessment of soil movement by manual tillage (Poesen *et al.*, 2000). Gerlarch troughs with hinged lid and measuring 0.6 m long, 0.4 m wide and 0.5 m high were installed in a shallow trench at the bottom of each field (Figure 3). The troughs were left open during land preparation by manual hoeing. The soil material lost in the Gerlarch troughs during manual tillage was collected directly for weight determination under field and oven dry conditions. Soil translocation by this method was calculated using equation (3) (Turkelboom *et al.,* 1997).

$$T = M*DM/W \tag{3}$$

Where: T = Soil flux (kg/m/tillage pass),
M = mass of moist soil lost in the Gelarch troughs (kg),
DM = Correction factor (Soil oven dry weight as percentage of M)
W = observation width (m). The length of the Gelarch trough (0.6 m) was taken as observation width.

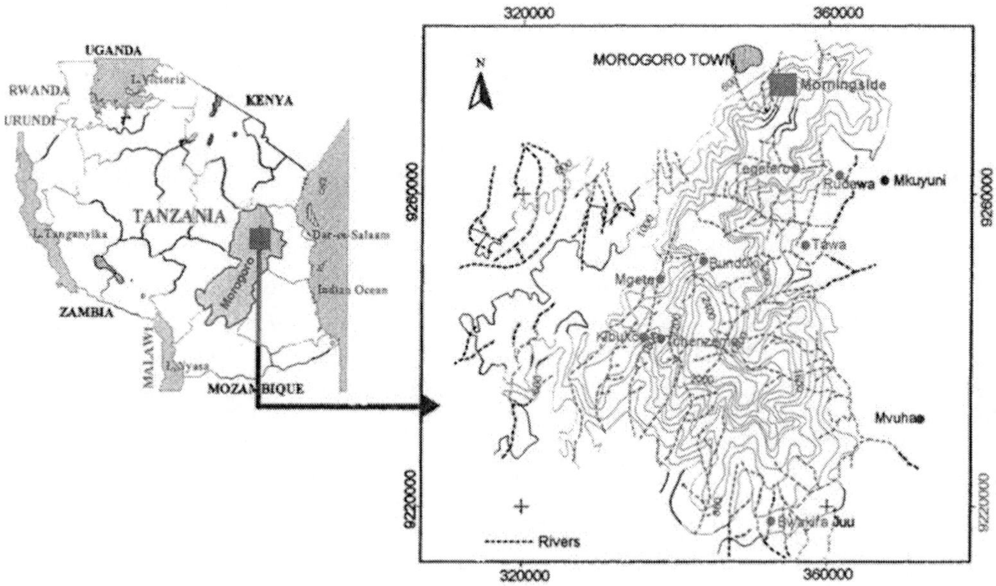

Figure 1. Location of the study area

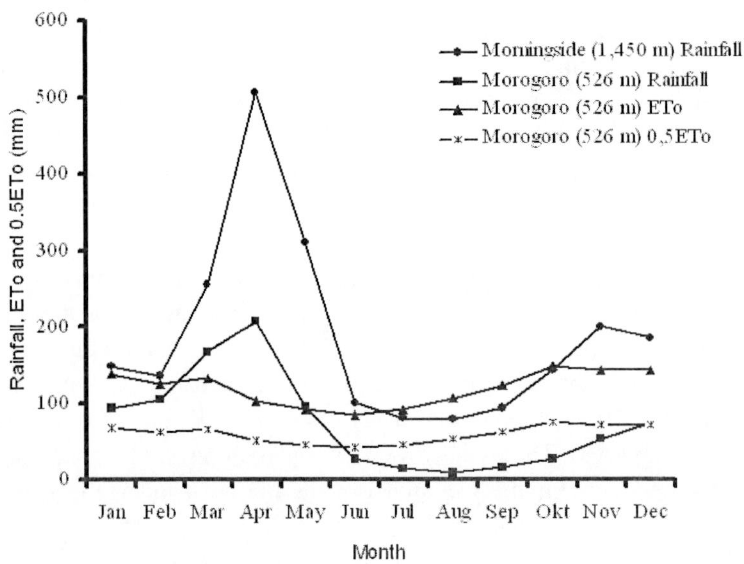

Figure 2. Mean monthly rainfall and reference evapotranspiration at Morningside and Morogoro climatic stations

Effect of soil moisture condition was studied on selected plots in areas under intensive agriculture at slopes 54 and 67 %. At each slope gradient soil flux was measured at three levels of soil moisture condition rated as low, intermediate and high for both shallow and deep tillage. The plots were irrigated using water cans until the plough layer was completely saturated. The irrigated plots were tilled to measure rates of soil flux for intermediate and high soil moisture conditions 96 and 48 hours respectively after application of irrigation. Soil flux with low soil moisture was measured after the plots were tilled at field moisture condition.

Figure 3. Measurement of soil translocation due to superficial manual tillage by Gerlarch trough method

Descriptive statistics (means, standard deviations, minimum and maximum) were widely employed in the exploratory analysis of soil flux rates and related factors. The effect of slope gradient and treatments were statistically analysed according to General Linear Model (GLM) of the Statistical Analysis System (SAS, 1999). Least Squares Means (LSMean) was used to compare the treatment means at 5 percent level of probability.

4.0. Correlation between Tillage Depth and Slope Gradient

Two types of manual tillage are practiced in the Uluguru Mountains; Superficial or shallow tillage (*kuparua* or *kuparaza*) and deep tillage (*kuchimba*). Superficial tillage is a common practice done by most farmers. In this practice a hoe ploughs only top few centimetres of the soil. A superficial tillage step is made after the soil surface is ploughed consecutively three or four times in either direction by the hoe blade at shallow depth and the soil material pulled backwards towards down slope direction. Mean tillage depth measured perpendicular to the soil surface in the study area varies between 3.5 cm and 6.5 cm. Average tillage depth was 5.2 cm (standard deviation = 0.7 cm, n=24). In the study area slope gradient was expected to be an important factor controlling the tillage depth but poor correlation (R^2 = 0.11, n = 24) was found between slope gradient and superficial tillage depth (Figure 4). Field observations made in northern Thailand indicated that tillage depth was influenced by many other factors such as shallow soils on steep slopes, rock fragments, texture and weed pressure (Poesen *et al.*, 2000).

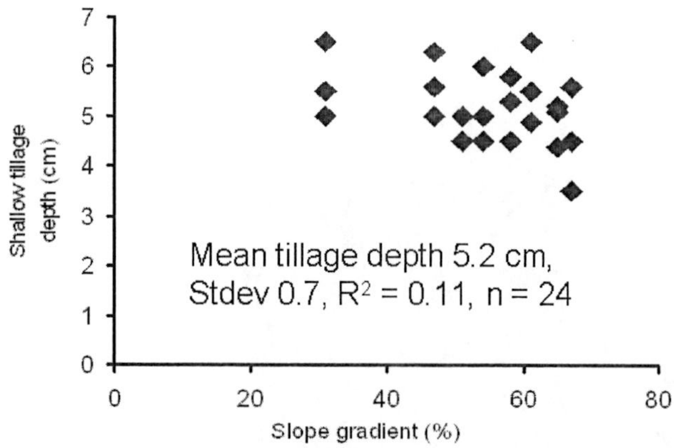

Figure 4. Relationship between slope gradient and shallow tillage depth perpendicular to the soil surface

Deep tillage is practiced only occasionally on fields which have been under fallow for relatively long period and are usually severely infested by weeds. It is also done when farmers are shifting from cultivation of one crop to another and also on fields with much grass in order to uproot the roots and rhizomes and to effectively destroy the roots by drying them on the soil surface. Mean tillage depth measured perpendicular to the soil surface for deep tillage ranges from 9.2 to 11.5 cm. Average depth was 10.4 cm (standard deviation = 1.6 cm, n=24). Deep tillage results in relatively big clods which started to roll or slide down freely by gravity on steep slopes. In the steep slopes of northern Thailand, a mean tillage depth perpendicular to the soil surface of 8.2 cm for manual deep tillage was measured (Turkelboom *et al.,* 1999).

5.0. Effect of Slope on Soil Flux

The mean soil flux rates due to manual tillage from different slopes are given in Table 1. The results showed that generally the rates of soil flux increased with increasing slope gradient. The mean soil flux increased significantly ($P<0.05$) from 16 kg/m/tillage pass at slope gradient 31% to more than 60 kg/m/tillage pass at slope gradient >60%. The higher rates of soil flux on slope gradient >60% could be attributed to the fact that the angle of repose for soil clods is exceeded whereby clods start to slide and roll down-slope by gravity. Turkelboom *et al.* (1999) observed in northern Thailand that there was modest increase in soil translocation on slopes ranging from 30 to 60% while on slopes steeper than 70% soil translocation increased significantly. Similar observations were also made by Van Muysen *et al.* (1999).

6.0. Effect of Tillage Depth and Soil Surface Condition on Soil Flux

The soil flux rates in relation to tillage depth and soil surface conditions (shallow and deep tillage with residues and bare surface conditions) are presented in Table 2. The rates of soil flux from bare soil surfaces for both shallow and deep tillage were higher than soil flux from soil surfaces with residues.

Table 1. Effect of slope on soil flux due to tillage by manual cultivation

Slope	Mean soil flux	Standard deviation	Minimum	Maximum	Number of Observations
%	Kg/m/tillage pass				
31	16^{a}	6	7	23	12
47	31^{b}	16	8	67	12
51	53^{c}	27	21	81	12
54	59^{dc}	30	21	92	12
58	59^{c}	21	35	95	12
61	67^{de}	24	34	101	12
65	75^{e}	23	40	100	12
67	70^{e}	27	44	111	12

Means with the same superscript letters are not significantly different (P<0.05)

Table 2. Effect of tillage depth and soil surface condition on soil flux due to manual tillage

Parameter	Mean soil flux	Standard deviation	Minimum	Maximum	Number of Observations
	Kg/m/tillage pass				
Shallow tillage with residues	31^{a}	3	7	47	24
Shallow tillage with bare soil surface	44^{b}	4	12	79	24
Deep tillage with residues	62^{c}	6	17	89	24
Deep tillage with soil surface	78^{d}	7	22	111	24

Means with the same superscript letters are not significantly different (P<0.05)

Mean soil flux of 31 and 62 kg/m/tillage pass were measured for shallow and deep tillage with residues compared to 44 and 78 kg/m/tillage pass measured on shallow and deep tillage with bare soil surfaces respectively. These differences were statistically significant (P<0.05) implying the contribution of soil cover in reducing soil flux.

The large differences observed in soil flux between shallow tillage and deep tillage could be explained by the nature of soil clods produced by the two tillage operations. Deep tillage produced larger and coherent clods with a diameter that ranged between 10 cm and 16 cm. These rolled down over a longer distance before they were completely destroyed. On the other hand shallow tillage produced clods with mean clod size of 2 to 5 cm which were smaller and relatively weak. Turkelboom (1999) observed soil flux rates of 39 to 87 kg/m/tillage pass for deep tillage with mean clod size ranging from 10 to 20 cm. Poesen *et al.* (2000) made similar observations in western part of northern Thailand. From this study it is clear that bare soil surface promotes tillage erosion. This is an important factor particularly in areas where farmers clear the fields and remove all the residues and/or set their fields on fire prior to cultivation. These practices should therefore be avoided in mountainous areas. A combination of both shallow tillage and presence of residues on the soil surface can cut down the rates of soil flux by 60%.

8.0. Influence of Both Tillage Methods and Surface Conditions on Soil Flux

The effects of both tillage methods and soil surface conditions are given in Table 3. Deep tillage on bare soil surface caused the highest soil flux which increased with slope gradient. The rates of soil flux due to deep tillage on bare soil surface ranges from 94 to 103 kg/m/tillage pass at slope gradient >60%. At a similar slope gradient, soil flux was significantly reduced (P<0.05) to lower rates ranging from 77 to 83 kg/m/tillage pass on the soil surface with residues (Table 3). Shallow tillage with residues minimised soil flux by a factor of nearly 2.5 within the slope range of 31 to 67%. Rates of soil flux were significantly less (P<0.05) at lower slopes (slope gradient <50%) and the same trends were observed between deep and shallow tillage.

Table 3. Mean soil flux due to manual tillage as influenced by tillage methods and soil surface conditions

Slope	Shallow tillage		Deep tillage	
	residues	bare	residues	bare
%				
31	9^{a}	14^{ab}	19^{ab}	22^{b}
47	20^{b}	23^{b}	28^{b}	54^{c}
51	24^{b}	36^{c}	72^{d}	79^{d}
54	28^{b}	39^{c}	82^{d}	86^{d}
58	39^{c}	53^{ef}	54^{e}	89^{d}
61	41^{c}	55^{e}	77^{f}	94^{dg}
65	43^{c}	77^{f}	83^{ef}	95^{g}
67	44^{c}	55^{e}	78^{f}	103^{g}

Means with the same superscript letters are not significantly different (P<0.05)

Residues were also effective in reducing soil flux irrespective of whether shallow or deep tillage was practiced but its effectiveness did not vary significantly with slope gradient notably under deep tillage. As for shallow tillage there was also a reduction of soil flux with decreased slope gradient irrespective of the soil surface conditions. Differences between bare soil surface and soil surface with residues were observed only at slope gradient >60%. The relatively large difference in soil flux between the bare soil surface and the soil surface with residues at higher slopes could be explained by the fact that the residues reduce the speed of clods and hence the reduction of soil fluxes. The effect of slope gradient on soil flux rates was more pronounced at slope gradient >60% for both shallow and deep tillage with bare and residue soil surface conditions respectively. The fact that the studied rates of soil flux seem to have similar behaviour at slope gradient <50% irrespective of the soil surface condition or the type of tillage depth being practiced seems to suggest that tillage erosion decreases when the slope gradient is less than 50%.

9.0. Effect of Tillage Method and Soil Moisture Conditions on Soil Flux

The differences on the rates of soil flux as influenced by tillage method and soil moisture conditions are presented in Table 4. The results showed that high soil moisture condition increased soil flux significantly ($P<0.05$) for both deep and shallow tillage. The observed high soil flux rates with increasing soil moisture condition could be explained by the easiness with which the tillage hand tools can penetrate the soil during tillage operation and hence deeper tillage depth by the farmers. The study revealed further that the amount of soil flux recorded under shallow tillage is less compared to deep tillage regardless of soil moisture condition. A maximum of 22 kg/m/tillage pass and a minimum of 15 kg/m/tillage pass were recorded for deep tillage while in the case of shallow tillage the corresponding values were 21 and 9 kg/m/tillage pass, respectively. As it was observed earlier deep tillage produced larger clods which rolled down over a longer distance before they were completely destroyed. Therefore on steep slopes deep tillage promotes tillage erosion regardless of soil moisture condition.

10.0. Conclusion

Rates of soil flux due to tillage erosion generally increased with increasing slope gradient. Bare soil surface promotes tillage erosion than soil surfaces with residues. It is also favoured by moist soil conditions. This is an important factor particularly in areas where farmers clear their fields and remove all the residues and/or set their fields on fire prior to cultivation. These practices should therefore be avoided in mountainous areas. A combination of both shallow tillage and presence of residues on the soil surface can cut down the rates of soil flux by 60%. The study showed that rates of soil flux increased with soil moisture content regardless of the type of tillage depth, a condition attributed to greater tillage depth.

Although it is not obvious to most farmers, tillage erosion is one of the most important processes of soil erosion that has to be taken into consideration in the overall setting of soil and water conservation priorities in the study area. Continued and extended studies on the determination of agro-ecological conditions influencing the frequency and depth of tillage and weeding are strongly recommended. The effect of tillage depth and displacement distance on soil flux rates due to superficial manual tillage should be investigated.

Table 4. Influence of tillage method and soil moisture condition on soil flux by manual tillage operation

Slope (m/m)	Moisture condition (%)	Tillage depth	Soil flux (kg/m/tillage pass)
0.54	8.4	deep	15^{b}
	16.2	deep	17^{b}
	24.4	deep	22^{a}
	8.4	shallow	10^{c}
	16.2	shallow	16^{b}
	24.4	shallow	11^{c}
0.67	8.6	deep	17^{b}
	16.1	deep	21^{a}
	22.1	deep	22^{a}
	8.6	shallow	9^{c}
	16.1	shallow	16^{b}
	22.1	shallow	21^{a}

Means with the same superscript letters are not significantly different ($P<0.05$)

ACKNOWLEDGMEMTS

The authors wish to express their sincere appreciation to the Soil and Water Conservation Project on the Uluguru Mountains under SUA-FLEMISH Inter-University Council (SUA-V.L.I.R.) Collaboration Programme for the financial assistance towards this work. The authors are grateful to the farmers of Mlimani Ward on the northern slopes of the Uluguru Mountains, who patiently shared their knowledge and wisdom about their way of living in the mountains.

REFERENCES

Dercon, G. (2001). *Tillage erosion assessment in the Austro Ecuatoriano* PhD thesis, Catholic University of Leuven, Belgium. pp 212.

Food and Agriculture Organisation (FAO), International Soil Reference and Information Centre (ISRIC) and International Society of Soil Science (ISSS) (1998). *World Reference Base for Soil Resources,* World Soil Resources Report, 84, FAO, Rome, Italy. pp 88.

Gerlarch, T. (1967). Hillslope troughs for measuring sediment movement. *Revised Geomorphology dynamics*, *42*, 197.

Govers, G., Quine, T., Desmet, P. & Walling, D. (1996). The relative contribution of soil tillage and overland flow erosion to soil redistribution on agricultural land. *Earth Surface Processes and Landforms*, *21*, 929-946.

Govers, G., Vandaele, K., Desmet, P., Poesen, J. & Bunte, K. (1994). The role of tillage in soil redistribution on hillslopes. *European Journal of Soil Science*, *45*, 469-478.

Guiresse, M. & Revel, J. C. (1995). Erosion due to cultivation of calcareous clay soils on the hillsides of southwest France: II Effect of ploughing down the steepest slope. *Soil and Tillage Research, 35*, 157-166.

Kachanoski, R. G., Miller, M. H. & Lobb, D. A. (1992). Soil loss by tillage erosion: The effects of tillage implement, slope gradient and tillage direction on soil translocation by tillage. A report prepared for the Soil and Water Environmental Enhancement Programme (SWEEP); Technology Evaluation and Development (TED) subprogram. Final Report. pp57.

Kiburys, B. (1995). Mechanical soil-erosion caused by tillage on 5°, 10° and 15° slopes. *Zemes ukio mokslai*, *4*, 10-13.

Kimaro, D. N., Deckers, J. A., Poesen, J., Kilasara, M. & Msanya, B. M. (2005). *Short and medium term assessment of tillage erosion in the Uluguru Mountains, Tanzania. Soil & Tillage Research*, *81*, 97-108.

Kimaro, D. N., Kilasara, M., Noah, S. G., Donald, G., Kajiru, K. & Deckers, J. A. (1999). Characteristics and management of soils located on specific landform units in the northern slopes of Uluguru Mountains, Tanzania. In: *Agricultural Research Challenges for the 21st Century. Proceedings of the Fourth Annual Research Conference of the Faculty of Agriculture*. (Edited by Faculty of Agriculture (FoA)) Sokoine University of Agriculture, Morogoro, Tanzania, November 17th-19th 1999, pp 234-242.

Lindstrom, M. J., Nelson, W. W. & Schumachter, T. E. (1992). Quantifying tillage erosion rates due to mouldboard plowing. *Soil and Tillage Research, 24*, 243-255.

Lindstrom, M. J., Nelson, W. W., Schumachter, T. E. & Lemme, G. D. (1990). Soil movement by tillage as affected by slope. *Soil and Tillage Research, 17*, 255-264.

Nyssen, J. (2001). *Erosion processes and soil conservationin a tropical moutain catchment under threat of anthropogenic desertification – a case study from northern Ethiopia* PhD thesis, Catholic University of Leuven, Belgium, pp 379.

Nyssen, J. (2001). *Erosion processes and soil conservationin a tropical moutain catchment under threat of anthropogenic desertification – a case study from northern Ethiopia*. PhD thesis, Catholic University of Leuven, Belgium, pp. 379.

Nyssen, J., Poesen, J., Mitiku, H., Moeyersons, J. & Deckers, J. (2000). Tillage erosion on slopes with soil conservation structures in the Ethiopian Highlands. *Soil and Tillage Research, 57*, 115-127.

Poesen, J., Turkelboom, F., Ohler, I., Somchai, Ongprasert, A. S. & Vlassak, K. (2000). Tillage erosion in northern Thailand: intensities and implications. *Mededelingen Des Seances*, *46(2000-4)*, 489-512.

Poesen, J., Van Wesemael, B, Govers, G., Martinez-Fernandez, J., Desmet, P., Vandaele, K., Quine, T. & Degraer, G. (1997). Patterns of rock fragment cover generated by tillage eroson. *Geomorphology*, *18*, 183-197.

Sampson, D. N. & Wright, A. E. (1964). The Geology of the Uluguru Mountains. *Geological Survey of Tanzania, Bullettin*, *37*, 1-69.

SAS, (1999). Statistical Analysis System, SAS/STAT User's Guide release. SAS Institute Inc, Carry NC, USA.

Sharma, A. K. (1987). Qualitative and quantitative land evaluation for rainfed maize in sub-humid tropical and sub-tropical climate. In: *Proceedings of the International Workshop on quantified Land Evaluation Procedures held in Washington, D C, 27 April - 2 May*

1986. (Edited by Beek, K. J., Burrough, P. A. & McCormack, D.E.) ITC publication No. 6. ITC, Enschede, The Netherlands, pp. 147-146.

Tsara, M., Gerontidis, S., Marathianou, M. & Kosmas, C. (2001). The long-term effect of tillage on soil displacement of hilly areas used for growing wheat in Greece. *Soil Use and Management*, *17*, 113-120.

Turkelboom, F. (1999). *On-farm diagnosis of steepland erosion in northern Thailand: Integrating spatial scales with household strategies*. PhD thesis, Catholic University of Leuven, Belgium. pp 309.

Turkelboom, F., Poesen, J., Ohler, I. & Ongprasert, S. (1999). Reassessment of tillage erosion rates by manual tillage on steep slopes in northern Thailand. *Catena*, *29*, 29-44.

Turkelboom, F., Poesen, J., Ohler, I., Van Keer, K., Ongprasert, S. & Vlassak, K. (1997). Assessment of tillage erosion rates on steep slopes in northern Thailand. *Soil and Tillage Research, 51*, 245-259.

Turkelboom, F., Poesen, J., Ohler, I., Van Keer, K., Ongprasert, S. & Vlassak, K. (1997). Assessment of tillage erosion rates on steep slopes in Northern Thailand. *Soil and Tillage Research, 51*, 245-259.

Van Muysen, W., Govers, G., Bergkamp, G., Roxo, M. & Poesen, J. (1999). Measurement and modelling of the effects of initial soil conditions and slope gradient on soil translocation by tillage. *Soil and Tillage Research*, *51*, 3003-316.

Van Muysen, W., Govers, G., Bergkamp, G., Roxo, M. & Poesen, J. (1999). Measurement and modelling of the effects of initial soil conditions and slope gradient on soil translocation by tillage. *Soil and Tillage Research*, *51*, 3003-316.

Zachar, D. (1982). *Soil erosion development in soil science*. Elselvier Scientific Publishing Company Amsterdam, Oxford, New York, pp. 547.

In: No-Till Farming
Editor: Earl T. Nardali

ISBN: 978-1-60741-402-5

Chapter 10

SURFACE STRATIFICATION OF SOIL NUTRIENTS IN NO TILL LIMITS NUTRIENT AVAILABILITY AND REDUCES GRAIN YIELD

***B. J. Radford*[A] *and B. A. Cowie*[B]**

[A]Queensland Department of Environment and Resource Management, LMB 1, Biloela, 4715 Queensland, Australia.

[B]Queensland Department of Environment and Resource Management, PO Box 1762, Rockhampton, 4700 Queensland, Australia.

ABSTRACT

Research has shown that the less mobile elements in the soil, such as P, K and Zn, become stratified in the surface 50 mm of soil after several years of continuous no-till cropping. This causes nutritional constraints to productivity when the surface soil becomes dry unless the rest of the soil profile is highly fertile or receives appropriate levels of fertiliser.

No-till crops grown in the semi-arid, subtropical environment of central Queensland, Australia, are particularly prone to nutritional disorders as a result of surface stratification. The surface soil can remain dry for long periods during crop growth while the roots obtain water stored in the heavy clay soils during the preceding fallow. Consequently nutrient deficiency symptoms have appeared despite apparently satisfactory levels of those same nutrient elements in the 0-100 mm layer, which is the standard sampling depth used to assess soil fertility.

The original aim of the 13-year experiment reported here was to assess the effect of tillage frequency and intensity, and stubble retention and removal, on soil water storage, soil nutrient status, and the growth and yield of rainfed grain sorghum in central Queensland. During the first 7 years of research, the mean yields of no till and traditional tillage (disc plough and scarifier) without fertiliser application were not significantly different. As the experiment continued, it became apparent that supplementary nutrition was required. The placement of appropriate levels of fertiliser elements (N, P, K, S and Zn) 100 mm deep and 50 mm to the side of each row resulted in large yield responses to no till. The yield of no till in comparison with traditional tillage (both with stubble retained and appropriate fertilisation) was 147%, 129% and 174% in the final three crops. These responses also reflect the outstanding potential of longer-term no till that has been demonstrated in other research work in central Queensland. Since mixing of the surface soil (by tillage) gave similar responses, it was concluded that some crop nutrients had become concentrated in the dry surface soil and were therefore inaccessible to plant roots during dry periods.

Keywords: no till, grain yield, surface stratification, soil nutrients

INTRODUCTION

A number of studies have found that the less mobile elements in the soil, such as P, K and Zn, become concentrated in the surface 50 mm of soil after several years of continuous no till cropping [Robson & Taylor, 1987; Cowie *et al.*, 1996]. This causes nutritional constraints to productivity when the surface soil becomes dry unless the rest of the soil profile is highly fertile or receives appropriate levels of fertiliser.

Crops grown in the semi-arid, subtropical environment of central Queensland, Australia, are particularly prone to nutritional disorders as a result of surface stratification. The surface soil can remain dry for long periods during crop growth while the roots obtain water stored in the heavy clay soils during the preceding fallow. This has resulted in nutrient deficiency symptoms after only 5-7 years of no till [Cowie *et al.*, 1996] despite apparently satisfactory levels of those same nutrient elements in the 0-100 mm layer, which is the standard sampling depth used to assess soil fertility. The high yield responses associated with no till in central Queensland [Radford & Thornton, 2009] encourage the practice of continuous no till and increase the rate of nutrient export. The nutrient most susceptible to stratification is phosphorus [Wright *et al.*, 2007].

The experiment reported here was planned at the start of the tillage revolution in Australia [Triplett & Dick, 2008]. Its original aim was to assess the effect of tillage frequency and intensity and stubble retention and removal on soil water storage, soil nutrient status, and the growth and yield of rainfed grain sorghum. The first seven years of research were reported in detail in the scientific literature [Thomas *et al.*, 1990 a, b; Standley *et al.*, 1990]. The mean yields of no till, blade plough/rod weeder tillage and disc plough/scarifier tillage were not significantly different during those first seven years when no fertiliser was applied. As the experiment continued, it became apparent that supplementary nutrients were required. When the appropriate elements (N, P, K, S and Zn) were applied at appropriate rates, there were large positive yield responses in the no-till treatment. A single tillage operation in the no-till treatment prior to sowing in order to mix the surface soil and redistribute the soil nutrients also gave a large positive yield response.

Materials and Methods

Site

The experiment was located on a property at Mt Murchison near Biloela (24.32°S, 150.53°E) in eastern central Queensland, Australia. The land originally supported a forest dominated by brigalow (*Acacia harpophylla*) before it was cleared between 1930 and1937. It is known to have been cropped annually since 1962, and no fertiliser was applied until December 1985.

The soil type is a grey Vertisol classified as an Entic Pellustert [Soil Survey Staff, 1975]. The clay content increases down the profile from 41% (0-100 mm) to 53% (140-160 mm). Details of some soil properties prior to the start of the experiment are given in Thomas *et al.* [1990a]. Land slope is about 3%.

The 84-year mean annual rainfall is 675 mm (range 317-1132 mm), with 72% falling in summer (October-March). The mean annual evaporation from a class A pan is 1870 mm.

Design and Treatments

The experimental design was originally a randomised block with 6 treatments and 4 replicates. The 6 treatments were disc, blade and no tillage (D, B and N) x stubble retained and stubble removed (+ and -). Disc and blade tillage were carried out as primary tillage operations, and a scarifier was used for all secondary tillage. Disc/scarifier tillage was the conventional fallow management technique when the experiment began in 1978. Blade/rod weeder tillage is a form of reduced tillage designed to reduce soil disturbance and stubble incorporation. In the no till treatment, all weed control was carried out using herbicides, and the only soil disturbance was caused by the sowing tines. Row spacing for the sorghum crops was 1 m except in 1990 when it was 0.75 m. The cotton was sown in twin rows 1 m apart with 3 m between the twins. The total number of tillage operations in both the disc and blade treatments was 47, an average of about four per fallow. The total number of spraying operations in the no-till treatment was 51, also an average of about 4 per fallow. Plots measured 25 x 15 m, and small diversion banks were constructed along the sides of the plots to prevent water running off one plot to the next. Water from the diversion banks drained into grassed laneways at the ends of the plots. After 7 sorghum crops, the plots were split for fertiliser application (control and fertilised), and a further 5 crops (4 sorghum and 1 cotton) were grown. The same tillage x stubble treatments were always maintained on the same plots, and the same fertiliser treatments on the same split plots, each year.

Table 1 shows the details of the cropping sequence, the amounts of fallow and in-crop rainfall received and the amounts of fertiliser elements applied to the plots that were split for fertiliser application.

In 1991, the final year of the experiment, the unfertilised subplots in 3 reps were split into 20 smaller subplots to apply various combinations of nutrient elements and two soil inversion treatments. The aim was to determine the level of response to five nutrient elements and to mixing of the surface soil to overcome surface stratification. Nutrients were applied in the following amounts: N (75 kg/ha), P (40 kg/ha), K (50 kg/ha), Zn (5 kg/ha) and S (30 kg/ha).

This fertiliser was banded 100 mm deep and 50 mm to the side of each row. Soil inversion was done before sowing with a single pass of a Rotavator set to a depth of 120 mm. The 20 fertiliser and inversion treatments are listed in Table 2. The design incorporated single, double and triple omission of the five nutrient elements in order to determine what was limiting when all other nutrients were in ample supply.

Table 1. Details of the crops grown, rainfall received and fertiliser applied to the split plots.

Crop and year of harvest	Cultivar	Sowing date	Harvest date	Fallow		Crop		Fertiliser applied to split plots
				Period days	Rainfall mm	Period days	Rainfall mm	kg/ha
Sorghum 79	E57+	21.2.79	27.7.79	250	598	156	129	Nil
Sorghum 80	E57+	16.1.80	29.4.80	173	274	104	109	Nil
Sorghum 81	E57+	9.1.81	11.5.81	255	231	132	366	Nil
Sorghum 82	E57+	30.12.81	29.4.82	233	293	121	266	Nil
Sorghum 83	E57+	19.1.83	19.5.83	265	277	120	543	Nil
Sorghum 84	E57+	11.1.84	3.5.84	237	336	113	128	Nil
Sorghum 85	E57+	19.12.84	2.5.85	230	421	134	377	Nil
Sorghum 86	E57+	19.12.85	16.4.86	231	462	118	241	50N + 10P
Sorghum 87	E57+	16.12.86	7.5.87	244	470	142	324	38N + 10P + 37K
Sorghum 88	E57+	9.2.88	10.6.88	278	378	122	320	75N + 40P + 82K + 39S
Cotton 90	Siokra	1.11.89	12.3.90	509	921	131	127	24N + 44P + 82K + 42S + 5Zn
Sorghum 91	DK 44M	18.1.91	8.5.91	312	579	110	76	75N + 40P + 50K + 30S + 5Zn

Table 2. Combinations of nutrient element and inversion treatments in the 1991 sorghum crop.

Treatment	Groupings
Control	Control treatments
S	
SN	Triple nutrient omission treatments (selected)
SP	
SK	
SZn	
SNP	Double nutrient omission treatments (selected)
SNK	
SNZn	
SPK	
SPZn	
SKZn	
SNPK	Single nutrient omission treatments
SNPZn	
SNKZn	
SPKZn	
NPKZn	
SNPKZn	All 5 nutrients
Soil inversion	Inversion treatments
Inversion + SNPKZn	

Measurements

Soil samples to assess surface stratification were taken on 25 January 1991 from only the disc and no till treatments, each with stubble retained and without any history of applied fertiliser. Depth increments sampled were 0-25, 25-50, 50-100, 100-150 and 150-200 mm. The 0-25 mm sample was taken using a square-section hand trowel, and the deeper samples were taken through this depression using a 25 mm diameter coring tube with a machined tip. For both treatments, 10 samples of each depth increment were obtained from three replicates, and the 30 samples were bulked and thoroughly mixed. Subsamples were then taken for analysis for bicarbonate-extractable P, exchangeable K, DTPA-extractable Zn and pH by the methods of Rayment & Higginson [1992].

Soil macrofauna were counted in all plots in May 1990 using a spade sampling technique [Robertson & Simpson, 1988] to collect 10 samples (180 x 180 mm) to a depth of 150 mm in each plot. All soil fauna exceeding ~0.5 mm in length were recorded. Samples therefore included some larger mesofauna [Swift *et al.*, 1979].

Grain yields were measured by means of a small plot header and, in 1990, with a mechanical cotton picker. In 1991, the smaller plots were hand-harvested. Grain moisture contents of subsamples were measured with a moisture meter, and all yields were adjusted to 12% moisture content. Cotton lint yield was measured at ambient moisture content. The datum area was 4 x 20 m in each plot except for the cotton in 1990 (4 x 10 m) and the sorghum in 1991 (1.5 x 3 m). The datum area was reduced in 1991 because in this final year all plots were further subdivided to impose multiple tillage x fertiliser treatments [Asghar *et al.*, 1993].

RESULTS

Nutrient Levels in the Surface Soil

The concentration of bicarbonate-extractable P was higher in N+ than D+ in the surface soil (0-25 and 25-50 mm) but levels did not vary with tillage practice at greater depths (50-200 mm) (Figure 1). Exchangeable K was higher in N+ at 0-25 mm but correspondingly lower in N+ at 50-100 mm. Extractable Zn was higher in N+ at 0-25 and 25-50 mm but lower in N+ at 50-100 and 100-150 mm. Soil pH was lower in N+ than D+ in the surface soil (0-25 and 25-50 mm) while values at greater depths were similar. These changes in the properties of the surface soil under no till resulted in marked deficiency symptoms in the cotton crop after 12 years of treatment application (Figure 2). Deficiency symptoms were also observed in the blade tillage treatments but to a lesser degree.

Soil Macrofauna

The population density of macrofauna in the soil after 12 years was significantly ($P<0.01$) higher in N+ than any other treatment. There were 82 m^{-2} in N+, 8 in N-, 20 in B+, 11 in B-, 26 in D+ and 11 in D-. Isopods comprised 46% of the population in N+.

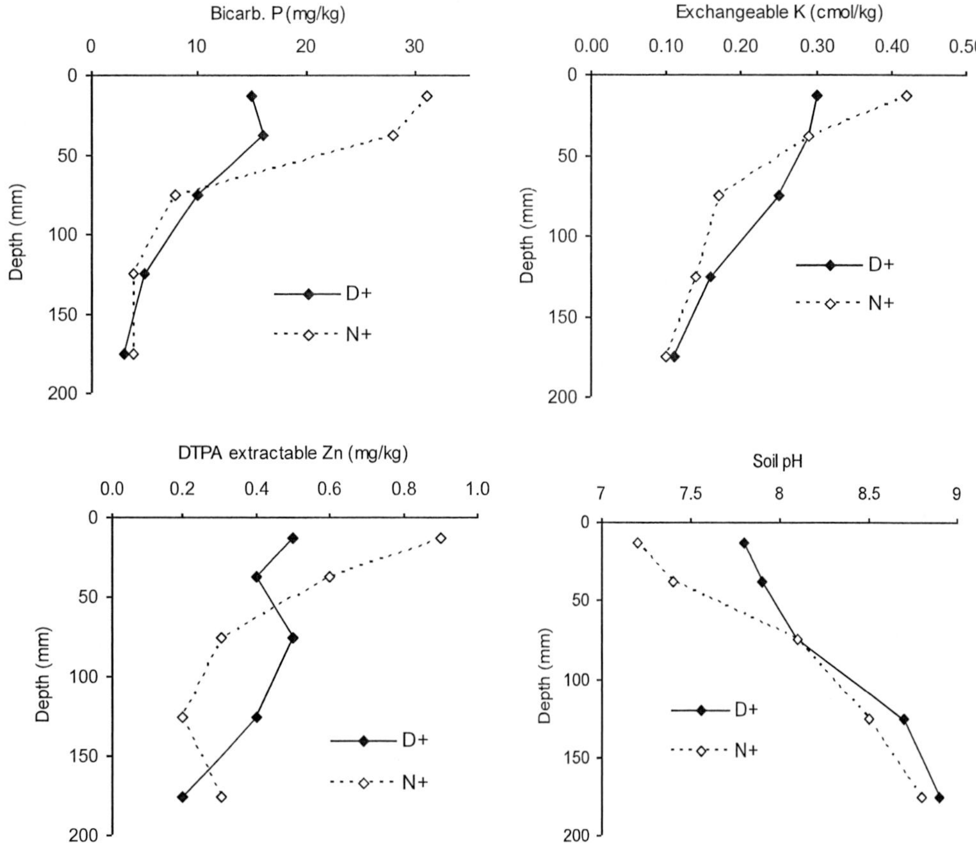

Figure 1. Effect of disc/scarifier tillage (D) and no till (N), both with stubble retained (+) on bicarbonate-extractable P, exchangeable K, DTPA-extractable Zn and soil pH in the surface soil in 1991.

Figure 2. Cotton crop at Mt Murchison after 12 years of no till with stubble removed, showing nutrient deficiency symptoms due to surface stratification (foreground) and the response to applied fertiliser (N, P, K, S and Zn).

Table 3. Statistically significant treatment differences in grain yield (t/ha) for 12 crops without applied fertiliser. Treatments are disc tillage (D), blade tillage (B) and no till (N) with stubble retained (+) and stubble removed (-). na: not available

Treatment	Crop number												Means
	1	2	3	4	5	6	7	8	9	10	11	12	
D -	3.32ab	1.40c	3.08bc	2.82cd	2.03c	2.65a	4.52abc	2.52b	1.30b	1.93b	0.22	1.09b	2.24
D+	3.61a	1.65b	3.59a	2.95bcd	2.26bc	2.97a	5.09a	3.00a	1.95a	2.13b	0.30	2.10a	2.63
B -	3.71a	1.68b	3.29ab	3.07abc	2.04c	2.77a	4.99ab	2.40b	1.61a	2.14b	0.25	2.44a	2.53
B+	3.74a	1.68b	3.57a	3.53ab	2.48b	2.83a	4.58abc	2.35b	1.52ab	2.69a	0.31	2.15a	2.62
N -	2.20c	1.55bc	2.77c	2.41d	1.74d	2.06b	3.71c	1.19c	0.51c	1.25c	0.04	1.14b	1.71
N+	2.90b	1.96a	3.12bc	3.64a	2.79a	2.86a	4.08bc	2.16b	1.43b	2.26b	0.21	2.27a	2.47
lsd (P=0.05)	0.60	0.25	0.41	0.64	0.28	0.45	0.92	0.45	0.45	0.36	na	1.04	na

Grain Yields

During the 13 years of the experiment, there was little difference in the mean yields of D+, B+ and N+ without applied fertiliser (Table 3 and Figure 3). No till yielded slightly less than the tilled treatments. In contrast, when adequate amounts of the appropriate nutrient elements were applied to the final three crops, N+ greatly outyielded D+ (Table 4 and Figure 3). The mean yield of N+ in the final three crops was half as much again as D+ (150%) while B+ yielded a quarter as much again (129%). These dramatic responses contrast sharply with the non-response without fertiliser, and indicate large yield gains are possible in continuous no till when appropriate crop nutrition is provided. N+ yielded 3.44 t/ha of sorghum in 1988 (D+: 2.34 t/ha) and 3.48 t/ha in 1991 (D+: 1.99 t/ha). N+ produced 511 kg/ha of cotton in 1990 while D+ gave only 396 kg/ha.

Table 4. Statistically significant treatment differences in grain yield (t/ha) for five crops with applied fertiliser. Treatments are disc tillage (D), blade tillage (B) and no till (N) with stubble retained (+) and stubble removed (-).

Treatment	Crop number					Means
	8	9	10	11	12	
D -	2.43c	1.55b	2.26b	0.37	2.24b	1.77
D+	3.10ab	2.07a	2.34b	0.40	1.99b	1.98
B -	2.50c	2.04a	2.31b	0.41	2.42b	1.94
B+	3.23a	2.09a	3.10a	0.41	2.99ab	2.36
N -	1.31d	1.25b	2.05b	0.42	2.07b	1.42
N+	2.69bc	2.24a	3.44a	0.51	3.48a	2.47
lsd (*P*=0.05)	0.45	0.45	0.37	na	1.04	na

na: not available

It should be emphasised that these responses to N+ occurred only when stubble was retained. When stubble was removed at the end of each season, the yield from no till was reduced, both with and without fertiliser application (Figure 4).

Yield responses to the different combinations of nutrient elements and to soil inversion are shown in Table 5. There was no significant (P>0.05) yield increase in response to the addition of any single nutrient.

The addition of P+K+Zn, however, always significantly (P<0.05) outyielded the control. If we compare the fully fertilised treatment with the five treatments lacking a single nutrient, yield was reduced only when P was removed. Where two elements were missing, yield was reduced by removing P+N, P+K or P+Zn (all three P treatments). Where three elements were missing, yield reductions again occurred only with the removal of the treatments containing P. Soil inversion resulted in a yield not significantly different from the complete fertiliser treatment.

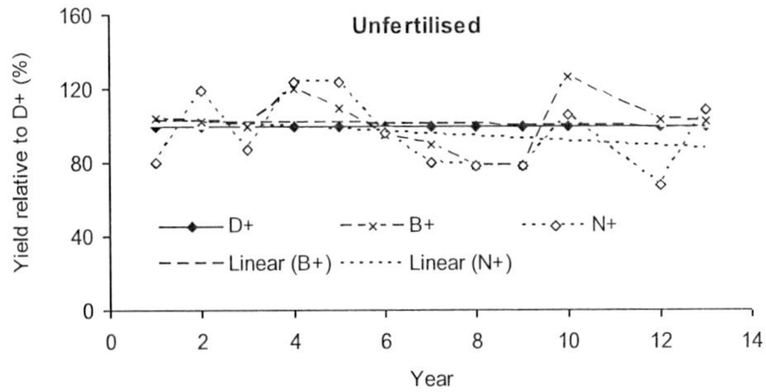

Figure 3. Continues on next page.

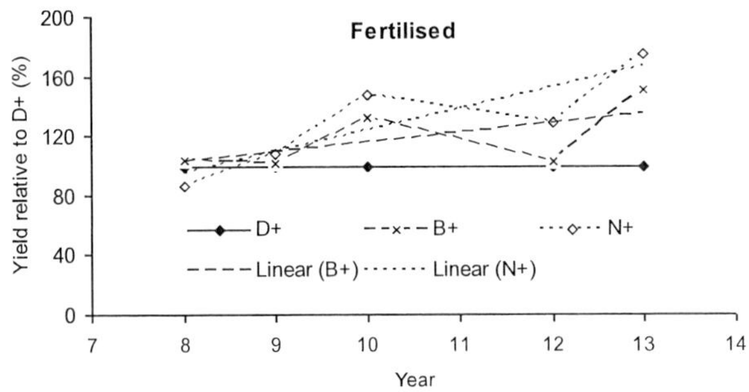

Figure 3. Grain yields of no tillage with stubble retained (N+) and blade tillage with stubble retained (B+) relative to traditional disc tillage with stubble retained (D+) for 12 unfertilised crops (top) and 5 fertilised crops.

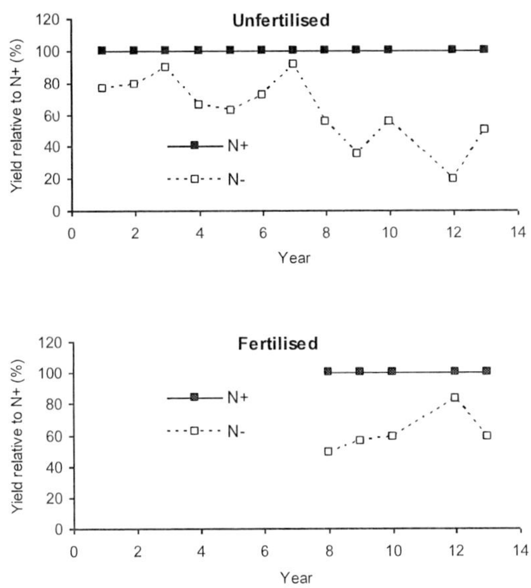

Figure 4. Effect of stubble retention (+) and removal (-) on crop yield under a no-till (N) regime with and without fertiliser application.

Table 5. Grain yield responses of 1991 sorghum to fertiliser and soil inversion treatments in no till with stubble retained (N+).

Grain yields in order of magnitude*	Treatment
3867 a	Inversion + SNPKZn
3623 ab	SPK
3475 abc	SNPKZn
3469 abc	SPKZn
3420 abc	NPKZn
2949 abcd	Inversion
2824 bcde	SNP
2747 bcde	SPZn
2665 bcde	SNPZn
2526 cdef	SP
2469 cdef	SNPK
2273 def	Control
2217 def	SKZn
2202 def	SK
2102 def	SNKZn
2082 def	SN
1959 def	SZn
1907 ef	S
1821 ef	SNZn
1531 f	SNK

* Values followed by the same letter are not significantly different ($P>0.05$)

Discussion

Stratification of relatively immobile soil nutrients (such as P, K, Zn, Cu, Mn, Fe and B) has been previously reported in response to no till [Robson & Taylor, 1987]. There were no associated yield reductions due to P, K or Zn stratification, however, probably because the soils were highly fertile, highly fertilised or moist enough at the surface to enable root access. Stratification problems appeared on brigalow soils in central Queensland after 7 years of no till at Mt Murchison and after only 5 years at Brigalow Research Station [Cowie *et al.*, 1996]. At both sites the surface soil (0-50 mm) was generally too dry for nutrient extraction by crop roots and levels of plant-available nutrients below 50 mm were low. At such sites the traditional 0-100 mm soil sample is inappropriate for assessing fertility because nutrients found at 0-50 mm are largely unavailable. A 50-150 mm sample would be more useful, as suggested by Cowie *et al.* [1996].

At a third site in central Queensland (Biloela), however, no stratification occurred after 7 years of no till [Cowie *et al.*, 1996]. Here soil mixing may have occurred during the passage of the sowing tines, when surface soil fell into soil cracks, and when earthworms and termites were active. The sowing tines were 1 m apart at Mt Murchison but 0.275 m apart at Biloela and Brigalow during the three years prior to sampling. Soil cracking would have been similar at all three sites because the clay content of their surface soil (0-100 mm) was similar (25-41%). At Biloela high earthworm population densities in no till (17 m^{-2}) [Wilson-Rummenie *et al.*, 1999] may have aided in redistributing the immobile elements [Darwin, 1881]. Termite activity under no till [Wilson-Rummenie *et al.*, 1999] may also have reversed stratification at Biloela.

Termites were also present at Mt Murchison [Holt *et al.*, 1993] but failed to overcome the stratification problem. A high total macrofauna population in N+ (86 m^{-2}) also failed to have any appreciable effect. This high population density of soil macrofauna, however, is a sign of superior soil health in N+ and may have enhanced soil structure, rainfall infiltration and water use efficiency.

Apart from the stratification problem, which can be overcome by adding nutrients, the condition of the N+ soil must have improved markedly after 10 years to produce the mean yield response reported here (150% of D+). Possible reasons include:

- Improved surface soil structure
- Reduced soil compaction
- Increased biological activity
- Increased soil organic matter and
- Increased plant-available water holding capacity through the leaching of salt below the crop root zone.

Tillage causes deterioration of soil physical properties due to the pulverising and smearing effects of implements, and pressure from load-bearing wheels, resulting in a marked reduction in infiltration capacity [Clarke, 1987]. Complete repair of the damage done by tillage implements, tractors and harvesting machinery takes several years. For example, soil compaction damage has persisted for 5 years under no till management [Radford *et al.*, 2007].

While compaction damage remains, soil macrofauna populations will also be low [Radford *et al.*, 2001] and will therefore contribute little to soil structural improvement. It has been shown that soil macrofauna take about 7 years to become abundant and diverse in temperate Australia after cultivation has ceased [Douglas, 1987].

Organic carbon at Mt Murchison declined at a slower rate in N+ than the other five treatments during the first 7 years of the experiment [Standley *et al.*, 1990]. This indicates that differences may have become larger after 10-13 years (when yield differences also became larger).

Before the experiment commenced, the levels of soil chloride below a depth of one metre were high: 811 mg kg^{-1} at 1.0-1.2 m, 991 at 1.2-1.4 m and 1090 at 1.4-1.6 m [Thomas *et al.*, 1990a]. Rayment & Bruce [1984] classed soil Cl levels of 600-2000 mg/kg as high. There were no notable changes during the first 3-5 years of the experiment [Standley *et al.*, 1990] but leaching may have occurred in subsequent years as rainfall infiltration into N+ improved. Any reduction in the soil Cl levels at 1.0-1.6 m depth would have increased the plant-available water content of the soil by allowing increased water extraction by plant roots [Dang *et al.*, 2008].

Our large responses to long-term (>10 years) no till have been duplicated in another experiment in central Queensland [Radford & Thornton, 2009]. Research in northern China has shown that continuous long-term (15 years) no till improved soil structure and biological activity [Li *et al.*, 2007]. This included significantly increased aggregate stability in the larger size classes, and greater capillary porosity (pores of diameter <60 μm). Canadian research showed that 20 years of continuous no till with full stubble retention but no nitrogen application outyielded 2 years of no-till; the 20-year no till produced 2.8 t/ha and the 2-year no-till 1.8 t/ha [Derpsch, 2008]. Increase in soil carbon content was considered the main factor leading to this result. Benefits from no till accrue over time, and Sá [2004] has recognised 4 phases in the evolution of a long-term no-till system: initial (0-5 years), transition (5-10 years), consolidation (10-20 years) and maintenance (>20 years). It is only in the maintenance phase that maximum soil benefits (physical, chemical and biological) are achieved.

Derpsch [2008] claims that any tillage performed during the earlier phases of no till means a return to the initial phase. If this is true, tillage to invert the soil and make surface-stranded nutrients available to crop roots merely fixes one problem and creates other longer-lasting ones. Although tillage makes nutrients in the surface soil more available, it also disrupts the network of vesicular arbuscular mycorrhizhae in no till [Kabir, 2004], which enhances nutrient uptake from the deeper soil.

This experiment has now been terminated but archived soil and plant samples can be made available to interested researchers for further analyses.

Conclusion

Surface stratification of nutrient elements with low soil mobility occurs after several years of continuous no-till management. Concentration of these elements in the surface soil can result in crop nutrient deficiencies when the surface soil is dry, causing marked reductions in yields. The elements implicated were P, K and Zn. The problem can be overcome by applying the deficient elements in fertiliser banded 100 mm deep and 50 mm to the side of each row. It can also be overcome with a single tillage operation to invert the soil, but tillage can do more harm than good in the long term. When crop nutrition was adequate, long-term (>10 years) no till with stubble retained gave large yield responses compared with traditional tillage with stubble retained. Yield increased by 47%, 29% and 74% in years 10, 12 and 13, respectively.

Acknowledgments

We thank Greg Thomas, Adrian Webb, John Standley and co-workers for instigating this work; Greg Thomas and Don Sinclair for leading the 1986 and 1987 experiments; Heather Hunter for leading the 1988 experiment; Les Robertson and his team for conducting the soil macrofauna counts; the late Mohammad Asghar for his small-plot work in the 1991 sorghum to ascertain the nutrients responsible for the yield response to no till; and Gary Blight, Melindie Hastie, Stephen Hunt, Mal Hunter and David Lack for their valuable assistance.

References

Asghar, M., Lack, D. W., Cowie, B. A. & Parker, J. C. (1993). Effects of surface soil mixing after long-term disc, blade and zero tillage on sorghum production in central Queensland. In: McDonald, G. K. & Bellotti, W. D. (Eds.), *Proceedings of the 7th Australian Agronomy Conference* (300-303). Parkville, Australia: The Australian Society of Agronomy, Inc.

Clarke, A. L. (1987). The need for fertility restoration/maintenance. In: Thompson, J. P. & Doughton, J. A. (Eds.), *Proceedings of the Queensland Crop Production Conference 1987* (7-17). Brisbane, Australia: QDPI Conference and Workshop Series QC 87007.

Cowie, B. A., Hastie, M., Hunt, S. B., Asghar, M. & Lack, D. W. (1996). Surface soil nutrient distribution following zero tillage and traditional tillage management. In: Asghar, M. (Ed.), *Proceedings of the 8th Australian Agronomy Conference* (160-163). Toowoomba, Australia: The Australian Society of Agronomy, Inc.

Dang, Y. P., Dalal, R. C., Mayer, D. G., McDonald, M., Routley, R., Schwenke, G. D., Buck, S. R., Daniells, I. G., Singh, D. K., Manning, W. & Ferguson, N. (2008). High subsoil chloride reduces soil water extraction and crop yield from Vertosols. *Australian Journal of Agricultural Research, 59,* 321-330.

Darwin, C. R. (1881). *The formation of vegetable mould through the action of worms, with observations on their habits.* London: John Murray.

Derpsch, R. (2008). No-tillage and conservation agriculture: a progress report. In: Godard, T. Zoebisch, M. A. Gan, Y. T. Ellis, W. Watson, A. & Sombatpanit, S. (Eds.), *No-till Farming Systems Special Publication No. 3* (7-39). Bangkok: World Association of Soil and Water Conservation.

Douglas, L. A. (1987). Effects of cultivation and pesticide use on soil biology. In: Cornish, P. S. & Pratley, J. E. (Eds.), *Tillage - New Directions in Australian Agriculture* (308-317). Brisbane, Australia: Inkata Press.

Holt, J. A., Robertson, L. N. & Radford, B. J. (1993). Effects of tillage and stubble treatments on termite activity in two central Queensland Vertosols. *Australian Journal of Soil Research, 31,* 311-317.

Kabir, Z. (2005). Tillage or no-tillage: impact on mycorrhizae. *Canadian Journal of Plant Science, 85,* 23-29.

Li, H. W., Gao, H. W., Wu, H. D., Li, W. Y., Wang, X. Y. & He, J. (2007). Effects of 15 years of conservation tillage on soil structure and productivity of wheat cultivation in northern China. *Australian Journal of Soil Research, 45,* 344-350.

Radford, B. J. & Thornton, C. M. (2009). Effects and after-effects of 20 years of reduced tillage practices on soil properties and crop performance in the semi-arid subtropics. In: Nardali, E. T. (Ed.), *No-Till Farming: Effects on Soil, Pros and Cons and Potential* (27-48). New York, USA: Nova Science Publishers, Inc.

Radford, B. J., Wilson-Rummenie, A. C., Simpson, G. B., Bell, K. L. & Ferguson, M. A. (2001). Compacted soil affects soil macrofauna populations in a semi-arid environment in central Queensland. *Soil Biology and Biochemistry, 33,* 1869-1872.

Radford, B. J., Yule, D. F., McGarry, D. & Playford, C. (2007). Amelioration of soil compaction can take 5 years on a Vertisol under no-till in the semi-arid subtropics. *Soil and Tillage Research, 97,* 249-255.

Rayment, G. E. & Bruce, R. C. (1984). *Soil testing and some soil test interpretations used by the Queensland Department of Primary Industries.* QDPI Information series QI84029. Brisbane, Australia: Queensland Department of Primary Industries.

Rayment, G. E. & Higginson, F. R. (1992). *Australian Laboratory Handbook of Soil and Water Chemical Methods.* Melbourne, Australia: Inkata Press.

Robertson, L. N. & Simpson, G. (1988). Sampling and dispersion of *Pterohelaeus alternatus* Pascoe and *Gonocephalum macleayi* (Blackburn) (Coleoptera: Tenebrionidae) larvae in soil. *Queensland Journal of Agricultural and Animal Sciences, 45,* 189-193.

Robson, A. D. & Taylor, A. C. (1987). In: Cornish, P. S. & Pratley, J. E. (Eds.), *Tillage - New Directions in Australian Agriculture* (284-307). Melbourne, Australia: Inkata Press.

Sá, J. C. M. (2004). Adubãcao Fosfatadano Sistema de Plantio Direto. In: Yamada, T. Silvia, & R. S. Abdalla (Eds.), *Sympósio sobre Fósforo na Agricultura Brasileira, Anais* (201-222). Piracicaba, São Paulo: POTAFÓS.

Standley, J., Hunter, H. M., Thomas, G. A., Blight, G. W. & Webb, A. A. (1990). Tillage and crop residue management affect Vertisol properties and grain sorghum growth over seven years in the semi-arid sub-tropics. 2. Changes in soil properties. *Soil and Tillage Research, 17,* 367-388.

Swift, M. J., Heal, O. W. & Anderson, J. M. (1979). *Decomposition in terrestrial ecosystems. Blackwell*, Oxford.

Thomas, G. A., Standley, J., Webb, A. A., Blight, G. W. & Hunter, H. M. (1990a). Tillage and crop residue management affect Vertisol properties and grain sorghum growth over seven years in the semi-arid sub-tropics. 1. Crop growth, water use and nutrient balance. *Soil and Tillage Research, 17,* 181-197.

Thomas, G. A., Standley, J., Hunter, H. M., Blight, G. W. & Webb, A. A. (1990b). Tillage and crop residue management affect Vertisol properties and grain sorghum growth over seven years in the semi-arid sub-tropics. 3. Crop residue and soil water during fallow periods. *Soil and Tillage Research, 17,* 389-407.

Triplett, G. B. Jr. & Dick, W. A. (2008). No-tillage crop production: a revolution in agriculture. *Agronomy Journal, 100,* S-153–S-165.

Wilson-Rummenie, A. C., Radford, B. J., Robertson, L. N., Simpson, G. B. & Bell, K. L. (1999). Reduced tillage increases population density of soil macrofauna in a semiarid environment in central Queensland. *Environmental Entomology, 28,* 163-172.

Wright, A. L., Hons, F. M., Lemon, R. G., McFarland, M. L. & Nichols, R. L. (2007). Stratification of nutrients in soil for different tillage regimes and rotations. *Soil and Tillage Research, 96,* 19-27.

In: No-Till Farming
Editor: Earl T. Nardali

ISBN: 978-1-60741-402-5
2010 Nova Science Publishers, Inc.

Chapter 11

REMOTE SENSING TO MONITOR NITROGEN DEFICIENCIES IN NO-TILL CORNFIELDS

Jun Zhang[1*], Alfred M. Blackmer[2], Jason W. Ellsworth[3] and Shuxia Sun[1]

[1]Wright State University, 3640 Colonel Glenn Highway, Dayton, Ohio 45435 (USA)
[2]In memory, Department of Agronomy, Iowa State University, Ames, Iowa 50011 (USA)
[3]Wilbur Ellis Company, 150 Burlington St., Pasco, Washington 99301(USA)

ABSTRACT

Remote sensing of canopy reflectance has great potentials to help develop a new generation of nitrogen (N) recommendations in the Corn Belt. Much of the information can be gathered through trials conducted at many different sites and years. A more immediate problem with the acceptance of this technique is that there is no established methodology for quantitatively describing the accuracy and precision of remote sensing for assessing N stresses, transforming these assessments to fertilizer recommendations, and evaluating the benefits of using these recommendations. Although qualitative methods are adequate to show the potential of remote sensing and in-season applications of N, quantitative methods are needed to assess the benefits obtained in production agriculture. Much of our effort was directed toward conducting field experiments in 1998-1999 with an appropriate design, gathering and analyzing the data. The data analyses address the complex problems caused by having different scales of measurement in the spatially variable no-till cornfields. The unique ability of this approach to making many observations with reasonable low cost helps reduce errors of measurement and make estimates of fertilizer needs more accurate and practical. Results obtained from this study clearly support the conclusion that remote sensing offers more flexibility for improving N management during corn production in the Corn Belt. Remarkable agreement was

* Corresponding author: E-mail: junior_zhang@hotmail.com. Present address: CORE Program - Nutrition Research Services. 3223 N. Broad St., Suite 175, Philadelphia, Pennsylvania 19140-5007 (USA).

obtained between assessments of N status based on canopy reflectance and actual grain yield responses. The data collected can be used to quantitatively estimate N fertilizer needs in spatially variable fields and to convincingly document the benefits of using remote sensing to guide in-season fertilization in no-till cornfields under the similar condition.

INTRODUCTION

Uncertainty in estimates of optimal rates of N fertilization prompts corn producers to apply more fertilizer than is usually needed by the crop (Kyveryga et al., 2009; Zhang et al., 2008a). This uncertainty is unavoidable because estimates of an optimal rate of N fertilization are necessarily based on observed yield responses to a set of fertilizer rates in the past (Macy, 1936; Heady et al., 1955; Leclerg et al., 1962; Waugh et al., 1973; Pearce 1983; Nelson et al., 1985; Cerrato and Blackmer, 1990; Colwell, 1994). The underlying problem is that conditions prevailing where the fertilizer will be applied often differ substantially from prevailing conditions when observations were made in the past.

The uncertainty that is associated with calibrating diagnostic tools becomes most apparent when data from different sites are pooled to develop diagnoses that are reasonably robust across the range of conditions likely to be encountered (Zhang et al., 2008ab). Major causes of the uncertainty include effects of factors other than N sufficiency on yields, spatial variability in important factors within and among plots, changes in the form of relationships as supplies of N approach and exceed optimal, errors of measurement, and a confounding of these factors (Zhang et al., 2008ab). The uncertainty shows up as inability to precisely define a critical value that distinguishes the plants having below-optimal supplies of N from those having optimal supplies of N.

Remote sensing of canopies has been recognized as an effective tool for monitoring N deficiencies during the growth of corn (Bauer, 1975; Walburg et al., 1982; Chappelle et al., 1992; Blackmer et al., 1995; Gitelson et al., 1996; Bausch and Duke, 1996; Blackmer et al., 1996ab; Blackmer and White, 1998; Shanahan et al., 2001; Zhang et al., 2006). This tool offers the opportunity to base diagnoses of N deficiencies on observations from an essentially unlimited numbers of plants. Blackmer and White (1998) demonstrated that this characteristic of remote sensing makes it possible to characterize spatial patterns in crop responses to N in fields where various rates of N were applied in strips. These strips cross areas of soil were assumed to include variability with respect to factors influencing supplies of N and other factors affecting yields of plants. They also demonstrated that spatial patterns in crop response as revealed by remote sensing could be related to spatial patterns in crop yield response as measured by combines equipped with GPS receivers and "on-the-go" yield monitors.

Chapelle et al. (1992) introduced the concept of ratio analysis of reflectance spectra, whereas Blackmer et al. (1996b) suggested use of relative reflectance and a non-limited N reference plot to calibrate instrument to a field situation and to compensate for fluctuating light conditions. Studies described in this paper explore the potential of using remote sensing and yield monitoring in field-scale trials to develop calibration data that makes remote sensing a useful tool for diagnosing N deficiencies and estimating N fertilizer needs in production agriculture. This study is based on the assumption that new technologies make it practical to collect unprecedented numbers of observations within and among fields by cooperating with producers who have yield monitors on their combines. It is reasoned that large numbers of observations should make it possible to avoid many of the problems encountered in studies where experiments were designed to accommodate the practical limitations that were imposed by relatively high costs and insufficient observations at a few locations.

Most corn producers in the Corn Belt currently apply all N fertilizer before their crops emerge. Much of this is applied in the previous fall. Research during the past decade, however, indicates that substantial economic and environmental benefits would be obtained if application of part of this N were delayed until corn plants are about 20 cm tall. Such delays reduce losses of fertilizer between the time of application and the time it is most needed by the crop. The delays also avoid the problem that extra N must be applied to compensate for the N losses and the problem that yield-limiting deficiencies occur when the losses exceed expectations. A reasonable delay in N fertilizer application makes it possible to make site-specific adjustments for the effects of spring weather on net amounts of N made available from soil organic matter.

Materials and Methods

Field experiments were conducted in Iowa during 1998-1999. Site code, location and soil association for the 12 cornfields are included in Table 1. The fields were planted to corn after soybean (*Glycine max* L.) in late April or early May, and managed by the producers using their normal practice in no-till cornfields except for N fertilizer treatment. Population density varied from 65,000 to 79,000 plants ha^{-1}.

Table 1. Site number, location and soil association for the 12 cornfields included in this study.

Site	Year	County	Township	Range	Size in acre	Association
1	1998	Blackhawk	88N	14W	30	Kenyon-Clyde-Floyd
2	1998	Blackhawk	82N	28W	65	Clarion-Webster-Canisteo
3	1998	Buchana	83N	8W	20	Kenyon-Dinsdale
4	1998	Hamilton	82N	28W	30	Canisteo-Clarion-Nicollet
5	1999	Boone	84N	27W	70	Canisteo-Clarion-Nicollet
6	1999	Linn	83N	28W	40	Canisteo-Webster-Nicollet
7	1999	Boone	83N	30W	60	Clarion-Nicollet-Webster
8	1998	Boone	87N	7W	15	Kenyon-Clyde-Floyd
9	1999	Greene	95N	27W	20	Spicer-Fieldon-Coland
10	1999	Boone	84N	26W	120	Canisteo-Clarion-Nicollet
11	1999	Greene	83N	29W	75	Canisteo-Clarion-Nicollet
12	1999	Buchana	86N	24W	120	Brownton-Ottosen-Bode

The 12 sites were selected to represent a wide range in responses to fertilizer N treatment. Sites 1, 3, 4 and 8 received 56 kg N ha^{-1} of N fertilizer (in the form of urea ammonium nitrate solution, 28-32% N) that was dribbled on the soil surface shortly before planting in order to avoid extreme N deficiencies. Figure 1 demonstrates one of the four sites prior to receiving various rates of N fertilizer. Sites 9 and 12 received liquid manure prior to planting at the rates equivalent to 168-190 kg N ha^{-1}.

When plants were about 18 cm tall, two or three rates of fertilizer N were injected midway between every other row to the 15 cm soil depth. The four sites receiving a pre-planting fertilizer application were applied with 0, 56, or 112 kg N ha^{-1} at this stage. All other eight remaining sites were applied with fertilizer rates of 56, 112, or 168 kg N ha^{-1} approximately at the similar stage. The fertilizer was applied in 6-row (76 cm spacing) strips going the lengths of the fields.

Each treatment was replicated at least 4 times in a stratified block design. The number of blocks was determined by the size and availability of each field. It is assumed that responses are highly correlated within block and weakly correlated between blocks.

Aerial photographs for each field were taken by the Aerial Services, Inc. (Cedar Falls, IA) using a Jena LMK model 1000 (152 mm) Precision Lens Aerial Camera System with forward image motion compensation mounted on a gyro-stabilizer in an airplane flying from a height of 1000-1100 m in early July and late August. All aerial photographs were taken under similar camera settings.

Each photograph was digitized by using a Hewlett-Packard ScanJet 6100 scanner and geo-referenced by using IMAGINE 8.4 software (ERDAS, Inc., Atlanta, GA). The photographs were digitized in a scale that gave about 100,000 pixels per hectare. Control points for geo-referencing were 60 by 120 cm plywood targets painted white and placed at the corners of each field. The photographs were not adjusted or standardized in color.

Grain yields were measured at one-second intervals by using combines equipped with an AgLeader Yield Monitor 2000 (AgLeader Technologies, Inc., Ames, IA). Positions of the individual yield sampling points were recorded by using a Trimble AgGPS 132 receiver (Trimble Navigation Ltd., Sunnyvale, CA) mounted on a conventional combine. The yield and moisture data were downloaded from a PCMCIA card on the yield monitor onto a PC computer.

Each field was divided into 4.5 by 6.1 m grids equivalent to widths of the strip and the tier in order to have the same scales for yield monitor data and canopy reflectance. Each grid contained at least four points of yield monitor data. Yield response was defined as the difference of the mean yields between the paired grids within adjacent strips receiving lower and higher N rates. Relative reflectance was defined as the ratio of reflected radiation in the same paired grids as for yield response. Cumulative frequency distribution curves for the responses were generated accordingly. Relative reflectance in green wavelength was shown and discussed in this paper for simplicity.

Extensions of Image Analysis and Spatial Analyst with ArcView 3.2 were used to analyze grid values of canopy images, grid means of yields, and digitized image data. Statistical data analyses including frequency distribution and quantile calculation were performed using SAS 9.2 (SAS Institute Inc., Cary, NC). Presentation of data in tables and figures is simplified by recognizing that the sites are numbered in an order of decreasing yield response to the first incremental increase of N.

Figure 1. No-till cornfield before (top: June 1) and after (bottom: July 8) receiving various rates of N fertilizer.

RESULTS AND DISCUSSION

Aerial Images and Frequency Distribution of Canopy Reflectance

Color aerial images taken in July and August bore little influence of the background soil, but mixed effects of N treatments, soil properties and other factors still existed, which were observed as differences in the greenness of corn canopy. The differences due to added N occurred in identifiable strips that usually could be distinguished from the effects of other factors. Lighter green indicated more reflectance due to less chlorophyll activity, whereas darker green indicated greater absorbance of visible light by chlorophyll.

Cumulative frequency distribution curves of the observed relative reflectance in the July and August images are shown in Figure 2a and 2b, indicating color changes at the canopy level as a measure of corn responses to N fertilization across the 12 sites. Calculating relative reflectance from small paired-grids within areas of relatively uniform soil helps to separate the effects of spatial variability in soil characteristics from the effects of the added N. This technique also minimizes the effects of light intensity or view angle, camera settings, film type, and the ways of film process and digitization across sites.

Responses of Grain Yields to N Fertilizer and Frequency Distribution

Cumulative frequency distribution curves for observed yield responses to N fertilization for paired-grid comparisons within each of the 12 different fields included in this study are shown in Figure 2c. Each curve represents 265 to 752 observations of yield responses.

The curve for yield responses to the first increment of added N (Y_{112}-Y_{56}) usually falls to the right of the curve for the second N increment (Y_{168}-Y_{112}) because yield responses to N usually show diminishing responses to each additional increment of N. The median yield response occurs at a cumulative frequency of 0.50 on each curve. Distances between lines at median yield responses should be expected to decrease as sites become less responsive to N. The lines essentially coincide in situations where neither increase in N rates affected yields. It is noteworthy to see that the relative positions of the two curves corresponding to responses to the first and second N increments were reversed at Site 10 and Site 12, which might indicate negative effects of N application at higher rates (168 kg N ha^{-1} versus 112 or 56 kg N ha^{-1}).

The median offers a robust estimate of the yield response that is less likely to be influenced by extreme measurements or other unusual conditions at specific areas within the field. This is important because, as noted by Zhang et al. (2000, 2008a), random errors in estimates of yield response impose serious limitations on ability to evaluate the tools that are used to diagnose deficiencies of N during the growth of corn. In addition to errors that occur when measuring yields, variation in estimates of yield responses can be caused by differences among plots although they have been assumed to be similar in all ways except supplies of N for crop growth.

The importance of a given yield response can be assessed by recognizing that fertilizers are applied to increase profits for the producer. Most producers, for example, would not fertilize if the expected yield response were not enough to pay for the fertilization cost. At the prices of corn grain and N fertilizer for the Corn Belt when these studies were conducted, a yield response of 0.25 Mg was about enough to pay for fertilization at a rate of 56 kg N ha^{-1}. As indicated in Table 2, the median yield responses were great enough to pay for the first increment of N at 9 of the 12 sites and for the second increment of N only at 3 of the 12 sites.

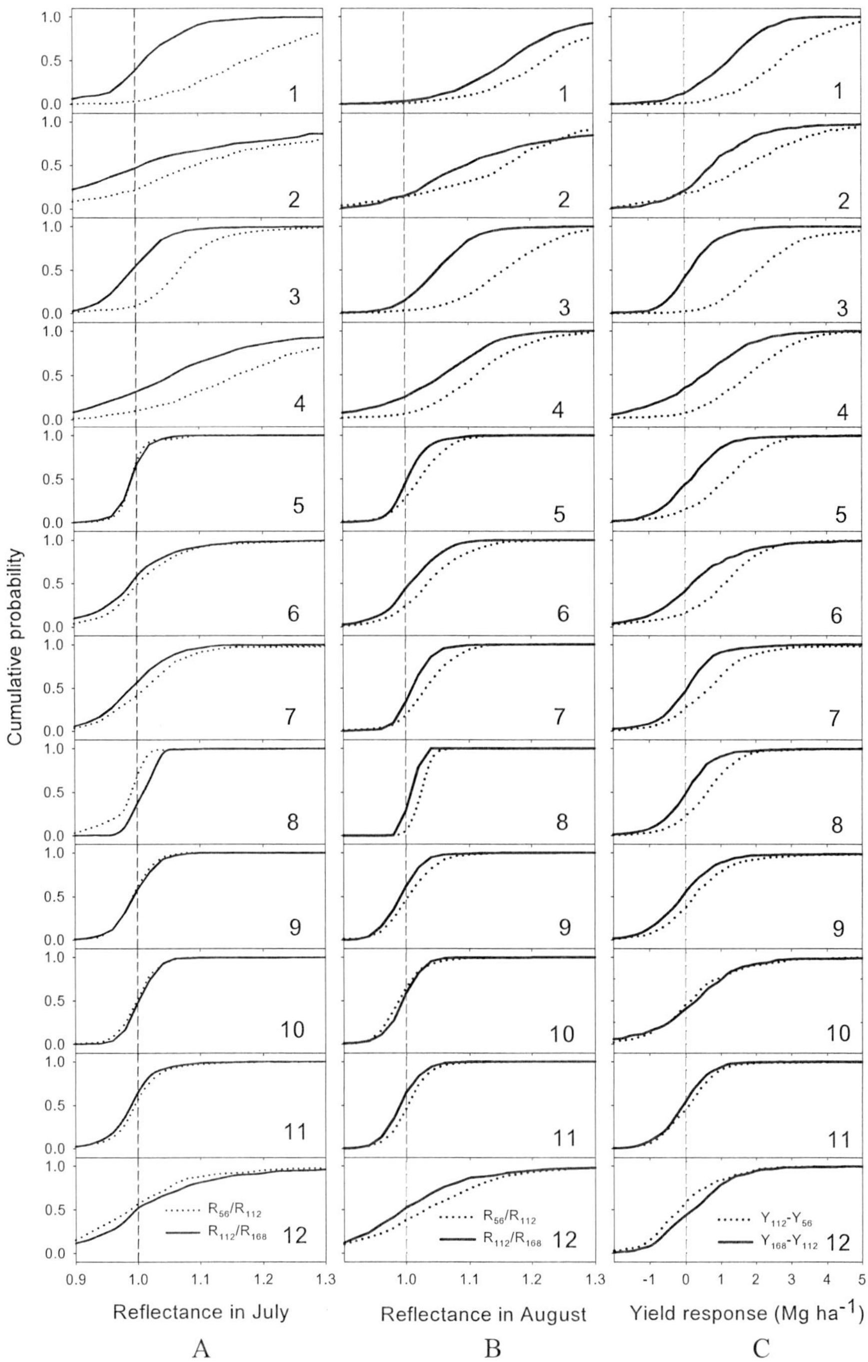

Figure 2. Cumulative probability for canopy reflectance measured in July and August as compared to grain yield responses to fertilizer N within each of the 12 fields. The vertical dashed lines indicate zero responses to fertilizer N.

Table 2. Medians of yield response and relative reflectance in two fertilization categories across 12 fields.

	Yield response ---------Mg ha^{-1}---------		Relative reflectance Measured in July		Measured in August	
Site	Y_{168}-Y_{112}	Y_{112}-Y_{56}	R_{168}/R_{112}	R_{112}/R_{56}	R_{168}/R_{112}	R_{112}/R_{56}
1	1.24	2.75	1.02	1.19	1.16	1.21
2	0.79	1.75	1.00	1.08	1.11	1.16
3	0.17	2.08	1.07	1.16	1.05	1.16
4	0.54	1.84	1.00	1.00	1.06	1.12
5	0.15	1.17	1.00	1.01	1.00	1.02
6	0.23	1.10	1.02	1.10	1.01	1.04
7	0.03	0.64	1.00	1.02	1.01	1.04
8	0.02	0.60	1.02	1.00	1.01	1.02
9	-0.01	0.25	1.01	1.01	0.99	1.00
10	0.16	0.24	1.01	1.00	0.98	0.99
11	-0.07	0.07	1.00	1.00	0.99	1.00
12	-0.13	0.21	1.00	1.00	1.00	1.04

Questions relating to the likelihood that an observed yield response within a field is more than random error are neither relevant nor important to the objectives of this study. Therefore, estimates of the "statistical significance" of yield responses would be influenced by amounts of within-field variability. However, decisions relating to whether or not to apply a given N increment would not be seriously influenced by this variability.

Relating Yield Responses to Responses in Relative Reflectance

Visual analysis revealed some agreement between the contrasts in color among strips in the images and analyses of yield responses presented in Table 2.

The relationships between median yield responses and median relative reflectance of the green band from the July and August images taken from each of the 12 experimental sites are shown in Figure 3. Observations for the first and second increments of added N are treated separately. Curvatures of the relationships are clearly shown in Figure 2, i.e. greater median yield responses at most experimental sites were not always associated with greater relative reflectance values. A modified two parameter power function $Y=a(1-X^{-b})$ was used to fit the non-linear relationships between yield response (Y) and reflectance response (X). These functions explained 64-72% of variations in yield responses to the first increment of added N and 49-63% of variations in yield responses to the second increment of added N across 12 experimental sites. These results indicate the noteworthy value of canopy reflectance measurements from the July images in the ability to diagnose relatively large shortages of N that likely occur earlier. However, smaller shortages of N usually would not occur in the midseason when a near optimal rate of fertilizer N had been applied early in the season.

Figure 4 shows the relationships between median yield responses and median relative reflectance of the green band from the images taken in July and August across 12 experimental sites. Linear equations were well fitted to a group of relative reflectance observations for the first and second increments of added N and explained 77-91% of the variation in yield responses to the added N. The August images seem more advantageous over the July images because both reflectance measurements and yield responses were made closer to the same maturity of corn and the effect of the N deficiency tends to accumulate over time.

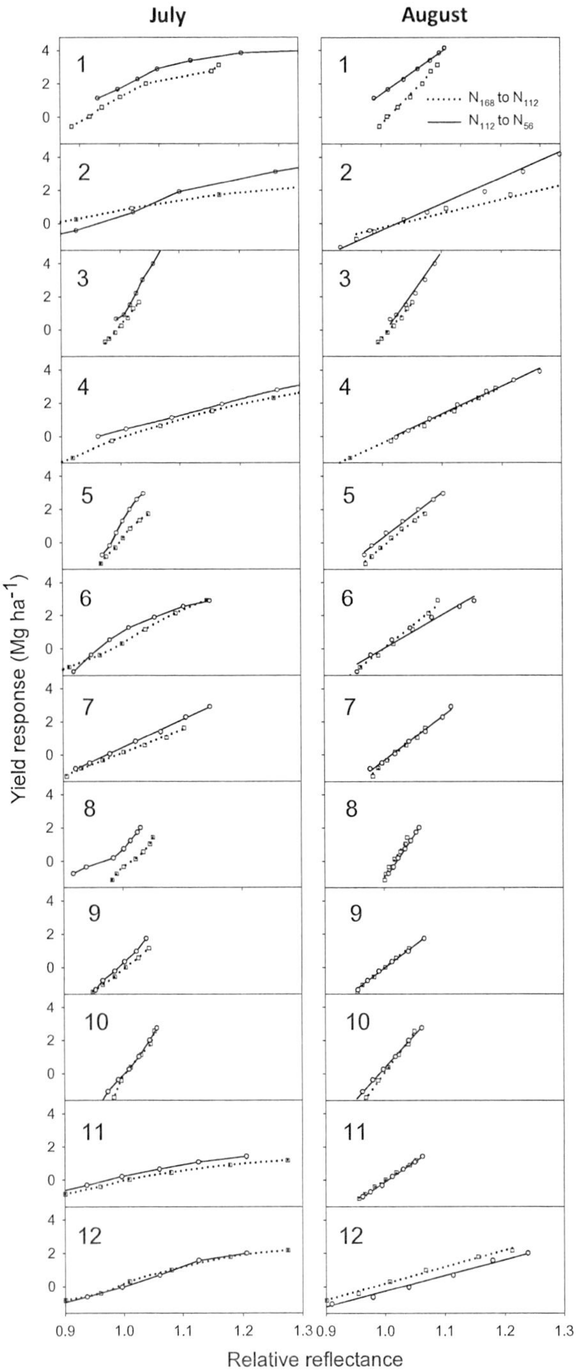

Figure 3. Relationships between quantiles of relative reflectance and quantiles of yield responses to fertilizer N within each of the 12 fields.

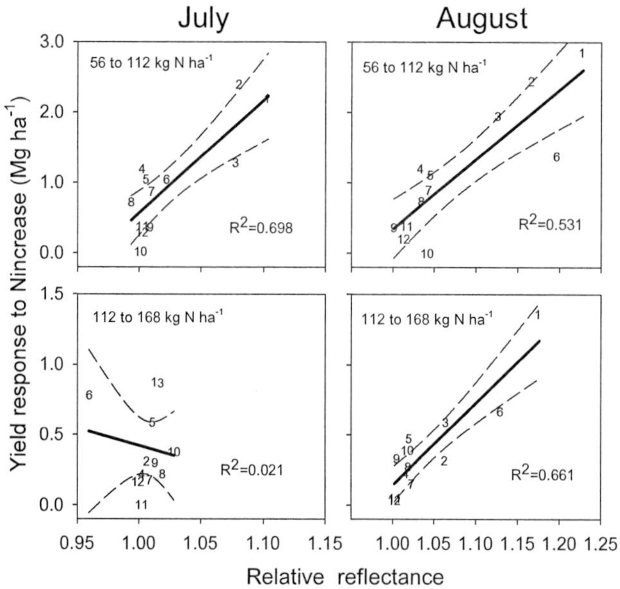

Figure 4. Relationships between medians of relative reflectance and medians of yield responses at different fertilization categories across 12 fields.

Discussion

Implications of Curved Relationships

The unexpected findings relating to curvature of the relationships between relative reflectance in July and final yield responses are consistent with observations indicating that the time of appearance of N deficiency symptoms depends primarily on how much supplies of N fall below optimal (Zhang et al., 2008b). The more severe shortages of N tend to cause symptoms to appear earlier. Plants growing on soils having near-optimal supplies of N often do not show any symptoms of N deficiency until late in the growing season. This situation should be expected if supplies of N in the soil were depleted during the rapid growth of corn.

Steep slopes at near-optimal supplies of N essentially indicate that midseason diagnosis did not reveal shortages of N whereas they were apparent at the end of the season when yields were measured. The relationships became more linear as the season progressed because both measurements were made closer to the same time. The effects of relatively large deficiencies became easier to detect as the season progressed because the effect of the N deficiency tends to accumulate over time.

This situation poses an ultimate limitation on the ability of midseason diagnoses to assess the need for small in-season applications of N fertilizer. The effects of relatively small shortages of N may not be expressed in plants until it is too late to correct the problem (Greenwood, 1976). It should be noted that our study explored the ability of remote sensing to detect the symptoms of N deficiencies that could be diagnosed when two different rates of N were applied early in the season. Assessments of fertilizer need would normally be based on observed responses to N applied after the diagnoses are made. This difference is important because there is little reason to believe that fertilizer N applied late in the season will have exactly the same effects on plant growth as the N fertilizer applied early in the season.

There seems to be noteworthy value in the ability to diagnose relatively large shortages of N by midseason diagnosis. Recent studies suggest that early season losses of N are often larger than generally recognized and that these losses can cause deficiencies of N that severely limit yields and profits (Balkcom et al., 2003).

Data collected later in the season could detect smaller shortages of N. Ability to detect small deficiencies only in the late season does not pose problems if the in-season observations are made primarily to learn how to reduce uncertainty in assessments of fertilizer needs for future years. With this use of remote sensing, the unique value of remote sensing resides in the ability to characterize spatial and temporal patterns in the occurrence of N deficiency symptoms across many fields at relatively little cost. Although yield monitors are needed in the studies to calibrate remote sensing for this purpose, remote sensing methods that have been appropriately calibrated can be used without the yield monitors.

Unique Value of Remote Sensing

The ability of remote sensing to make measurements on unprecedented numbers of locations within fields offers several important advantages. Some of these advantages occur when remote sensing is used with on-the-go yield monitors to establish the relationships that serve as the basis of the calibrations for making diagnoses in production agriculture. One obvious advantage is that diagnoses can be made over areas large enough to measure small differences in yield with a high degree of confidence. When focusing on rates very near optimal, whole fields can be covered with strips having different N rates and studied at negligible costs by cooperating with producers who normally harvest with yield monitors.

Another advantage of this approach is that taking the median of many observations minimizes problems associated with errors of measurement. Such errors could be within-plot sampling errors associated with making measurements on different plants, random errors associated with measurement of yield, or random errors caused by differences between plots due to factors other than N availability.

Calibrations based on observations made at many different sites and years provide an objective basis for estimating the reliability of the method for diagnosing N deficiencies in production agriculture. As noted in our previous paper (Zhang et al., 2008a), there is great need to question the assumption that data from controlled small plot experiments can be used to calibrate tools for diagnosing deficiencies in production agriculture. It seems likely that the mid-season assessments of N deficiencies are not commonly used because producers see no evidence that they are reliable.

Refinements in techniques are almost certain to substantially increase the ability of remote sensing to detect relatively small deficiencies of N during the growing season. It should be noted, however, that superior methods of diagnosing deficiencies can not be identified without using experimental methodology that distinguishes the superior from inferior methods. Superior methods will not be used in production agriculture unless producers can detect the benefits on their fields.

CONCLUSION

Remote sensing to assess deficiencies of N during the growth of corn seems to have unique potential because measurements can be made on a scale that enables reliable assessments of performance as the technique is normally used in production agriculture. A great strength of this approach is that it allows collection of very large numbers of observations at relatively little cost. Observations across a large number of fields and years establish robust relationships that can be used to estimate fertilizer needs. Remote sensing methods that focus on the current sufficiency of N for growth are more likely to be successful than methods that focus on measuring yield.

The greatest limitation on the use of remote sensing for this purpose seems to be in the lack of established methodology for doing agronomic studies on scales that express the unique advantage of remote sensing. Refinements in the methods we used are likely to greatly increase ability to diagnose mild N deficiencies during the growth of corn.

REFERENCES

Balkcom, K. S., Blackmer, A. M., Hansen, D. J., Morris, T. F. & Mallarino, A. P. (2003). Testing soils and cornstalks to evaluate nitrogen management on the watershed scale. *J. Environ. Qual.*, *32*, 1015-1024.

Bauer, M. E. (1975). The role of remote sensing in determining the distribution and yield of crops. *Adv. in Agron.*, *Vol.27*, 1975.

Bausch, W. C. & Duke, H. R. (1996). Remote sensing of plant nitrogen status in corn. *Trans. of the ASAE. 39*, 1869-1875.

Blackmer A. M. & White, S. E. (1998). Using precision farming technologies to improve management of soil and fetiliser nitrogen. *Aust. J. Agric. Res.*, *49*, 555-564.

Blackmer, T. M., Schepers, J. S. & Meyer, G. E. (1995). Remote sensing to detect nitrogen deficiency in corn. In *Site-Specific Management for Agricultural System*, 505-512.

Blackmer, T. M., Schepers, J. S. Varvel, G. E. & Walter-Shea, E. A. (1996a). Nitrogen deficiency detection using reflected shortwave radiation from irrigated corn canopies. *Agron. J.*, *88*, 1-5.

Blackmer, T. M., Schepers, J. S. Varvel, G. E. & Meyer, G. E. (1996b). Analysis of aerial photography for nitrogen stress within corn fields. *Agron. J.*, *88*, 729-733.

Cerrato, M. E. & Blackmer, A. M. (1990). Comparison of models for describing corn yield response to nitrogen fertilizer. *Agron. J.*, *82*, 138-142.

Chappelle, E. W., Kim, M. S. & McMurtrey III, J. E. (1992). Ratio analysis of reflectance spectra (RARS): An algorithm for the remote estimation of the concentrations of Chlorophyll A, Chlorophyll B and carotenoids of soybean leaves. *Remote Sens. Environ.*, *50*, 161-170.

Colwell, J. D. (1994). Estimating fertilizer requirements: A quantitative approach. CAB International, Wallingford, Oxon, UK.

Gitelson, A. A., Kaufman, Y. J. & Merzlyak, M. N. (1996). Use of a green channel in remote sensing of global vegetation from EOSMODIL. *Remote Sens. Environ.*, *58*, 289-298.

Greenwood, E. A. N. (1976). Nitrogen stress in plants. *Advance in Agron.*, *28*, 1-35.

Heady, E. O., Pesek, J. T. & Brown, W. G. (1955). Crop response surfaces and economic optima in fertilizer use. *Iowa Agri. Exp. Station Research Bull.*, 424. Ames.

Kyveryga, P. M., Blackmer, A. M. & Zhang, J. (2009). Characterizing field-scale variability in corn yield response to nitrogen fertilization. *Agron. J.*, *101(2)*, 269-277.

LeClerg, F. L., Leonard, W. H. & Clark, A. G. (1962). Field plot technique. Burgess Publishing Company.

Macy, P. (1936). The quantitative mineral nutrient requirements of plants. *Plant Physiol*, *11*, 749-764.

Nelson, L. A., Voss, R. E. & Pesek, J. (1985). Agronomic and statistical evaluation of fertilizer response. p. 53-90. *In* O.P. Engelstad (ed.) *Fertilizer technology and use*. SSSA, Madison, WI.

Pearce, S. C. (1983). The agricultural field experiment: *A statistical examination of theory and practice*. John Wiley & Sons.

Shanahan, J. F, Schepers, J. S. Francis, D. D. Varsel, G. E. Wilhelm, W. W. Tringe, J. M. Schlemmer, M. R. & Major, D. J. (2001). Use of remote-sensing imagery to estimate corn grain yield. *Agron. J.*, *93*, 583-589.

Walburg, G., Bauer, M. E. Daughtry, C. S. T. & Housley, T. L. (1982). Effects of nitrogen nutrition on the growth, yield, and reflectance characteristics of corn canopies. *Agron. J.*, *74*, 677-683.

Waugh, D. L., Cate, R. B. & Nelson, Jr. L. A. (1973). Discontinuous models for rapid correlation, interpretation, and utilization of soil analysis and fertilizer response data. North Carolina State Univ. *Tech. Bull*, No. 7.

Zhang, J., Blackmer, A. M. & Ellsworth, J. W. (2000). Interrelating corn canopy reflectance and yield response to fertilizer N applied in strips. p. 288. In: 2000 *ASA Abstr.* ASA, Madison, WI.

Zhang, J., Blackmer, A. M., Kyveryga, P. M., Glady, M. J. & Blackmer, T. M. (2006). Quantifying temporal patterns in symptoms of nitrogen deficiencies in corn by remote sensing. *In 2006 ASA Abstr.* Madison, WI. http://a-c-s.confex.com/crops/2006am/techprogram/P26842.HTM

Zhang, J., Blackmer, A. M., Kyveryga, P. M., Glady, M. J., & Blackmer, T. M. (2010). Temporal patterns in symptoms of nitrogen deficiency as revealed by remote sensing of corn canopy. *Pedosphere. 20 (1),* 1-10.

Zhang, J., Blackmer, A. M., Ellsworth, J. W. & Koehler, K. J. (2008a). Sensitivity of chlorophyll meters for diagnosing nitrogen deficiencies of corn in production agriculture. *Agron. J., 100,* 543–550.

Zhang, J., Blackmer, A. M., Ellsworth, J. W., Kyveryga, P. M. & Blackmer, T. M. (2008b). Luxury production of leaf chlorophyll and mid-season recovery of nitrogen deficiencies in corn. *Agron. J., 100*, 658–664.

INDEX

A

B

C

F

G

H

P

U

V

W

Y

Z

T